Severe Storm Forecasting

Tim Vasquez

2009

SEVERE STORM FORECASTING
First edition

January 2010
November 2015 color edition

Copyright ©2010 Tim Vasquez
All rights reserved

For information about permission to reproduce selections from this book, write to Weather Graphics Technologies, P.O. Box 450211, Garland TX 75045 or servicedesk@weathergraphics.com. No part of this publication may be reproduced, stored in a retrieval system, or transmitted by any means without the express written permission of the publisher.

ISBN 978-0-9969423-0-0
Printed in the United States of America

table of contents

The Forecast Process   1
1. A brief history   3
2. The forecast process   5
3. Objective methods   6
4. Subjective methods   10
5. The balanced forecast   12
6. Scales of motion   14
7. Wind expressions   17

The Thunderstorm   25
1. Storm morphology   27
2. The updraft   29
3. Mesocyclone   32
4. Downdraft   33
5. Storm motion   37
6. Splitting storms   39
7. The severe storm   41
8. Multicell storms   42
9. Supercell thunderstorms   43
10. Heavy rain and flooding   51
11. Cloud structures   53

Mesoscale Convective Systems   57
1. MCS types   59
2. MCS features   61
3. MCS dynamics   68
4. Bow echo storms   70
5. Derecho   72
6. MCS propagation   74

Tornadoes   77
1. Supercell tornado lifecycle   80
2. Supercellular tornado mechanisms   83
3. QLCS tornadoes   88
4. Miscellaneous tornado types   89
5. Tornado damage   91
6. Tornado prediction   93
7. Tornado nowcasting   96

Hail   99
1. Hail formation   101
2. Hail prediction   105
3. Hail detection   108

Lightning   109
1. Lightning types   111
2. Lightning formation   113
3. Dipole characteristics   118
4. Lightning detection systems   119
5. Lightning forecasting   121

Stability & Shear   123
1. Types of instability   125
2. Sounding basics   127
3. The skew-T log-p diagram   127
4. Lifting a parcel   130
5. Parcel methods   132
6. Modification   134
7. Sounding proximity   136
8. Sounding characteristics   136
9. Thermodynamic diagnostics   141
10. The hodograph   144
11. Instability-shear relationships   153
12. Moist symmetric instability (MSI)   154
13. Composite parameters   157

Radar   159
1. Base products   161
2. Limitations of radar   163
3. Polarimetric radar   169
4. Polarimetric base products   171
5. Polarimetric radar derived products   174
6. Dual doppler radar   175
7. Storm evolution principles   175
8. Severe storm signatures   176
9. Hail signatures   180
10. Tornado signatures   184
11. Other valuable signatures   187

Satellite   189
1. Image types   191
2. Cloud forms   194
3. Storm signatures   198

Diagnosis   201
1. Upper level patterns   203
2. Moisture   207
3. Pressure   211
4. Fronts   213
5. Other boundaries   216
6. Dryline   217
7. Subjective diagnosis   220
8. Surface target zones   221
9. Convective modes   224
10. Special characteristics   227
11. Localized phenomena   227
12. Other phenomena   231
13. Dynamic forecasts   232

APPENDIX   239
WSR-88D description   241
Diagnostic Variables   245
Hodograph   249
Skew-T log P Diagram   250
References   251
Index   261

preface

My interest in severe weather forecasting can be credited to Alan Moller and Skip Ely, who during my 1985 visits to the technical library at NWS Fort Worth always sent me home with the latest papers and journals on the emerging topics of mesoanalysis and storm structure. At that time, very little about the subject had diffused out to the public except through journals, and not one of the titles in my popular weather book collection even mentioned supercells except for a single National Geographic article. Sifting through the bounty of journal papers I felt like a brand new field that unlocked the mysteries of Texas weather had been laid before me. Seeking to make sense of it all, I began subscribing to AMS journals, practiced mesoanalysis and diagnosis almost daily, and by 1987 began storm chasing to travel the back roads of Texas and Oklahoma and witness the ingredients coming together.

My true test in mesoanalysis came on October 12, 1993 at Dyess Air Force Base, Texas. The ink had barely dried on my forecaster certification and it was my first dryline event. I went straight to the maps and data. A short while later, under clear skies, I issued a severe weather advisory, which prompted the command post to order B-1 bombers and C-130 transports into any available hangars. Two hours later skies darkened and later golf-ball sized hail rained down on the flightline. That event established my reputation at Dyess. It was not sheer luck or a stroke of genius, but rather a fortunate payoff from a habit of meticulous analysis and diagnosis. It goes to show that anyone with the dedication or experience can succeed, too.

Severe Storm Forecasting was a project started in 2007 to serve as the perfect companion for intermediate forecasters and a refresher for experienced forecasters. It maps the current state of severe thunderstorm forecasting from an operational framework rather than a research or academic perspective. Equations, physics, and lengthy citations have been kept to a bare minimum.

The meager number of references is not to discredit those who laid the groundwork but rather to keep the style as fluid and readable as possible. I have compensated by crediting key findings where possible and providing a lavish collection of references in the appendix which, unconventionally, contain comments and notes explaining the significance of the reference. Still, it was not possible for me to include every useful reference; a check of the AMS journal index shows an astounding 1300 abstracts that deal with severe weather, not counting conference preprints! The papers I have listed are the ones that focus most directly on fundamentals of forecast concerns: things like analysis, techniques, and conceptual models.

That said, if you have comments about this book, concerns about the matter presented, a demand for accuracy or clarification, or have suggestions for future editions, I would be glad to hear from you (send it to servicedesk@weathergraphics.com). Unlike the typical nonfiction title that hits the market, this is not a project that's kicked out to the printer and abandoned for other pursuits. Nearly all my books go into second and third editions which grow increasingly popular, so I fully expect there to be a revised edition in the next year or two. You may contact me through the e-mail address provided on the copyright notice page.

Special thanks to Les Lemon, Bob Johns, Jim LaDue, Rich Thompson, Paul Markowski, Carrie Langston, Roger Edwards, and Rich Rotunno for their assistance and comments. I also thank Olivier Staiger, Bill Hark, David Hoadley, and Paul Morley for use of their photos and artwork. Thanks to Kevin Crawmer, and L.B. LaForce for corrections. If there is anyone I have forgotten, rest assured it's inadvertent — let me know and I'll arrange an instream job correction with the printer. Most of all, special thanks to my wife Shannon Key for her patience during the many dusk-to-dawn nights I spent in November 2009 finishing the book.

TIM VASQUEZ
Norman, Oklahoma
December 2, 2009

notes

Due to the reversed Coriolis sign in the southern hemisphere, some concepts are marked as (NH) to indicate they apply in the northern hemisphere, with (SH) indicating the southern hemisphere. These are omitted where the hemisphere is already mentioned. Much of severe weather research, tradition, and history is bound to the northern hemisphere so the nature of habit may result in a few oversights throughout the text, but I hope the annotations are sufficient to help forecasters in Australia, South Africa, Argentina, Brazil, and other hot spots south of the equator.

The Master said, "If one learns from others but does not think, one will be bewildered. If on the other hand, one thinks but does not learn from others, one will be in peril."

- Analects II:15, c. 300 BC

1

THE FORECAST PROCESS

The scientist does not study nature because it is useful to do so. He studies it because he takes pleasure in it, and he takes pleasure in it because it is beautiful. If nature were not beautiful it would not be worth knowing, and life would not be worth living. I am not speaking, of course, of the beauty which strikes the senses, of the beauty of qualities and appearances. I am far from despising this, but it has nothing to do with science. What I mean is that more intimate beauty which comes from the harmonious order of its parts, and which a pure intelligence can grasp.

— Jules Henri Poincaré
Science and Method, 1908

Severe weather forecasting is perhaps unlike any other human endeavor, relying on highly nuanced aspects of scientific knowledge and intuition to anticipate an imminent catastrophe that has not yet even appeared. While for many others professions decisions are questions of shrewd resource management or understanding complex ranges of prescribed methods and correctly applying them, severe weather forecasting is a field where absolutely no single method of problem-solving is dependable. Those who rely upon a favorite technique or leave it all to numerical predictions soon entangle themselves in confusion and failure.

Even as a discipline within meteorology, severe weather forecasting stands out. "Ordinary forecasting" involves relatively rich data sampling, predictable forces which are largely in equilibrium, and a lengthy research history, while severe weather forecasting involves relatively coarse data, atmospheric motion largely out of balance, and only a few decades of rock-solid scientific progress. Consequently, severe weather forecasting success depends on an exceptional mix of analytic and intuitive thinking. Arguably only the most brilliant or dedicated forecaster can consistently find success in severe weather forecasting. And since a forecast success story never owes itself to a single model chart or a lucky guess, the importance of properly understanding the nature of the forecast process is absolutely imperative.

1. A brief history

Up through the Middle Ages, thunderstorms were widely regarded as a random scourge of nature. Only with the rise of science in the 19th century was it demonstrated that weather systems, and at a smaller scale, storms, had a fairly predictable nature. Telegraph technology had been deployed nationwide by the 1860s and during the Reconstruction years the U.S. government established a network of weather reporting stations.

U.S. Army officer John P. Finley combed through the first ten years of national weather maps and identified a number of weather patterns that were conducive to tornado weather. In 1891, the U.S. Weather Bureau was created and control of the observation and forecasting network became a regimented government function. Although technology was in place to allow for the growth of a severe storm alert network, this capability was largely ignored and

Storm forecasting in 1907

"We adopted the practice of being more conservative in our forecasts when we anticipate tornadoes. We do not use the term 'tornado' any longer. We say, 'Conditions are favorable for severe thunderstorms.' We stop there, because every tornado is accompanied by thunderstorms, and a tornado is probably nothing more or less than a thunderstorm stood on end. In a thunderstorm the air is rotating about a horizontal axis, like this. Coming toward you it rotates. In a tornado the axis is vertical.

"In the thunderstorm there will probably be little or no difference in temperature between the two ends of the horizontal rotating column. But turn it vertically and there is a vast difference in temperature between the bottom and the top. There is your energy between the two extremes. So that we predict thunderstorms for a region that we anticipate may have tornadoes; and we do so for the further reason that there will be such a small region struck by the tornado."

WILLIS L. MOORE
director, U.S. Weather Bureau
House committee hearing, 1907

the Weather Bureau chose to focus on frontal systems and hurricanes for many decades.

However during the first half of the 20th century, hundreds of lives were lost every year in tornadoes. The Tri-State outbreak in 1925 killed over 700 Americans. With an apparent rise in tornado incidence, mounting concerns over public safety and the danger to a rapidly expanding airline network, Congress forced the U.S. Weather Bureau in 1952 to begin addressing the severe weather forecast problem. This resulted in the creation of the Severe Local Storms Unit, or SELS.

Fortunately, firm groundwork had just been laid. The first detailed microscale examination of thunderstorms had been made in the Thunderstorm Project, headed by Horace Byers and Roscoe Braham. It analyzed thunderstorms in Florida in 1946 and Ohio in 1947 using new radar technology in combination with surface and upper-air data. The Air Force had also experienced success with internal tornado forecasts, particularly Robert C. Miller's 1948 warnings for the Tinker AFB command post. This along with operational experience at SELS led to dramatic advances in storm forecasting techniques during the 1950s. But very little was known about tornadic storms and they were assumed to be connected with the violent weather of squall lines.

During the late 1950s, University of Chicago researcher Theodore Fujita studied the Fargo tornado in great detail and established the first conceptual model for a tornadic thunderstorm. The widespread arrival of radar in the 1960s and 1970s and photography and observation from field experiments helped unravel storm structure and life cycle, leading to a model of the supercell in the late 1970s whose basic elements have not changed since.

At the same time, mesoscale meteorology, the study of subsynoptic weather, came into its own right as a specialized branch of meteorology. It specifically addressed non-hydrostatic processes, caused by any significant motion that overwhelms the vertical equilibrium in the atmosphere. The combination of further storm research, the emergence of tools like Doppler radar, and the study of mesoscale meteorology led to incredible advances during the 1980s and early 1990s that drastically improved the ability of forecasters. It is not practical to summarize them here as those findings are distributed throughout this entire book.

Though discoveries have come in rapid succession since the first harvests of the Thunderstorm Project, many small details are still not well-understood, and computer technology still limited to dynamical and analytic work. There is still substantial dependency on the subjective element of forecasting and the role of the human forecaster.

Figure 1-1 (top left). Ernest Fawbush, left, and Robert Miller, right, reviewing a forecast. In 1948 they put out the first official tornado warning, however it would be several years before the general public reaped the benefits. Note the use of teletype printouts and manual plots. *(US Air Force)*

Figure 1-2 (bottom left). Severe forecasting in the 21st century. Shown here is the National Weather Service forecast office in Fort Worth, where in this scene meteorologists are keeping tabs on a severe thunderstorm north of Abilene. *(Tim Vasquez)*

2. The forecast process

The goal of forecasting is to find out exactly what is happening now and how it is likely to change. The decision making process is threefold, requiring perception, comprehension, and projection of the information. In meteorology, this is known as analysis, diagnosis, and prognosis.

2.1. ANALYSIS ("WHAT"). The analysis process solves the question of *what* weather is taking place. This is the very first step in the forecast puzzle, a process of perception. The forecaster examines current observed weather as depicted by graphics and data, views remote-sensing data such as radar and satellite imagery, and even looks out the window. This procedure constructs a four-dimensional mental model of the current state of the atmosphere. Before

---

A clouded crystal ball

We think that in the long run, frankly, that [tornado watches] will not be very acceptable to the public.

FRANCIS W. REICHELDERFER
U.S. Weather Bureau director
March 1952

> **Stratospheric thinking**
>
> The decisionmaking process can be illustrated in this example from Air Force pilot Dan House, whose Mach 3 SR-71 jet experienced an emergency at 74,000 ft near the Philippines on 21 April 1989, as detailed in Col. Richard Graham's "SR-71: Stories, Tales, and Legends":
>
> ◆**Analysis**. Slight thump. Nose yaws to left. Engine RPM is zero. A and B hydraulic systems are losing pressure.
>
> ◆**Diagnosis**. The yaw and RPM loss are consistent with engine failure. A compressor explosion of some kind has damaged both hydraulic systems.
>
> ◆**Prognosis**. Damage to the hydraulic lines will lead to irreversible failure as fluid is lost. When the hydraulic pressure drops to zero, directional control will be lost.
>
> House and Bozek prepared accordingly for bailout and ejected successfully along a remote portion of the Philippine coast. They were rescued by fishermen and dressed in their "moon suits" were treated to a meal at a town mayor's house until rescue helicopters arrived. The SR-71 broke up after hitting the water and salvage efforts began immediately to prevent the reconnaissance equipment and wreckage from falling into Soviet hands.

the 1990s, the analysis process usually involved a shortage of information due to graphics and communication system limitations. The situation has turned upside down and forecasters now have an overabundance of graphics and data, requiring the forecaster to prioritize instead of improvise, and to have a working familiarity with all products that might be used so that the best one can be selected. Since graphical products such as surface charts have vastly more raw information than a forecaster can hope to assimilate, the technique of graphical hand analysis helps the forecaster methodically consider key quantities and find patterns.

2.2. DIAGNOSIS ("WHY"). The diagnosis process seeks to understand *why* observed weather is occurring. This is the process of comprehension. The forecaster synthesizes information from numerous kinds of graphics, images, and other products, and mentally develops a coherent, unified understanding of meteorological process taking place across the forecast area. If diagnosis has been accomplished, any item of interest noted in the analysis, such as a spot of cold temperature or the existence of a storm, can be correctly explained. This diagnosis is not a separate process in itself but grows and solidifies while the analysis takes place. When the diagnosis is complete, the forecaster can confidently answer *why* a certain aspect of observed weather is taking place.

2.3. PROGNOSIS ("HOW"). Prognosis process determines *how* observed weather will change. This is the process of projection. It's based upon the analysis and diagnosis and allows the forecaster's own intuition, insight, and experience to shape the forecast. A proper human prognosis will have firm expectations with regard to frontal and air mass movement, temperature trends, and weather events even when a computer model forecast is not available. In practice, the human prognosis process is augmented by computer models, particularly beyond 12 to 24 hours in the future, but conceptual models, experience, gut feeling, and of course understanding from the diagnosis significantly *shapes* the results.

3. Objective methods

A number of objective techniques exist for analyzing, diagnosing, and forecasting the weather. An objective technique is a rigid,

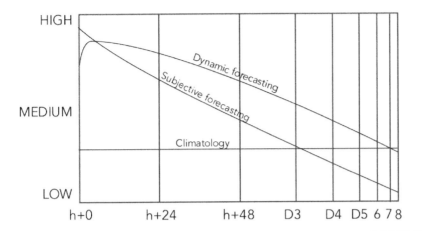

Figure 1-3. Forecast skill comparing dynamic forecasting (from numerical models) and subjective forecasting (the human forecaster without access to model data). What is most notable here is that the human excels in the first few hours, because the brain has a powerful capacity for diagnosing the current state of the atmosphere, and coupled with reason and intuition can make a fairly accurate prediction of changes a few hours into the future. The models are simply unable to resolve the fine-scale nature of the atmosphere at the level and detail offered by radar, satellite, and a glance out the window. This makes the severe storm forecaster's role exceptionally important.

However the equations of motion and physical parameterizations do become dominant once all the small-scale processes have played out, and beyond 4 to 6 hours a model will generally outperform a human forecaster. Several days into the future, the model offers only a slight advantage over climatological normals, and this domain is called medium-range forecasting. After a week into the future, dynamic forecasts are generally poor, though they may handle large-scale details fairly well. This prediction range is called long-range forecasting.

analytic scheme based on mathematical, meteorological, physical, and even statistical relationships. In short, a certain measure of skill is already built into the technique. As a result, an individual who knows little about meteorology can successfully apply an objective method to obtain a meaningful forecast result.

3.1. DYNAMICAL NUMERICAL MODELS. Dynamical numerical models are heavily used in severe weather forecasting and are known simply as "the models". They are essentially sophisticated three-dimensional simulations based on the equations of motion, allowing air motion, temperature, and humidity to be solved mathematically at any given point in time. Because the atmosphere cannot be precisely modelled, the errors in the model simulation grow substantially over time. By 5 to 10 days in the future, the results of most modern numerical models compare poorly to actual charts of the atmosphere at that forecast time, and errors only get worse as the forecast valid time goes farther out, 16 days being the maximum that most weather agencies will go.

Another effect of our inability to precisely model the atmosphere is the inability to precisely resolve important features which might help cause thunderstorm activity. In some cases, a difference of just one Celsius degree in the mid-levels of the atmosphere may be the difference between clear blue skies and a tornadic supercell. With only one or two radiosonde balloons taking measurements in each state every morning, numerical models are highly dependent on them. There are some techniques for working around this

Tornado forecasting

The advent of tornado forecasting is undoubtedly one of the most ambitious ventures into operational weather prediction since the inception of the science of meteorology. For years a "hands off" policy was deemed the only proper manner to treat this weather phenomenon.

R. WHITING & R. E. BAILEY
Eastern Air Lines, 1957

Figure 1-4. A flowchart example. Moller's flowchart for southern Plains storms was developed in the late 1970s and demonstrates things that Texas-based forecasters from that era looked for. It can be seen that this is a pattern-based forecast method which works entirely from features found on a surface chart. The nonlinearity of the atmosphere makes any rigid decision tree much too simplistic and myopic for forecast use, but such tools are valuable for demonstrating the interrelation of forecast elements. In this case Moller used the decision tree to illustrate the relation between moisture axes, thermal axes, and pressure falls.

limitation, but balloon observations are the backbone of numerical modelling. By contrast, humans are often able to extract details about the atmosphere between radiosonde sites simply by viewing satellite and radar imagery.

The numerical models used for routine forecasting have become quite good at solving the statewide or nationwide weather maps, but are rather poor at resolving individual thunderstorms. They are certainly not capable of modelling hail, tornadoes, and storm evolution. Thus numerical models are very useful for establishing the environment that prepares the storm, but are not useful once the storms are underway.

3.2. DIAGNOSTIC PARAMETERS. Forecasters may make use of diagnostic parameters, which simply express the observed data in the form of a forecast outcome. This does not even require the use of a computer or numerical model and can often be calculated by hand or estimated. A well-known example of a diagnostic parameter is a stability index or shear index. A diagnostic parameter offers a simple method and a clear forecast outcome.

While diagnostic parameters can help simplify forecast tasks, they can present a number of problems. First, they are a *proxy* for

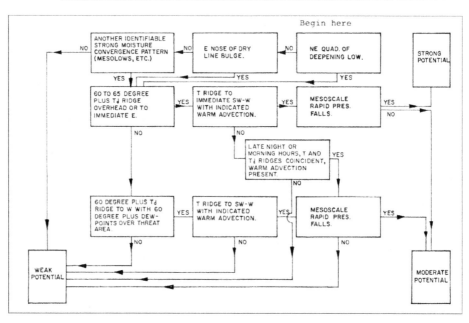

the actual underlying meteorological processes taking place and only represent it through abstraction. A parameter may be *contextually inappropriate*, such as when surface-based instability fields are examined north of a warm front. Another problem is *unrepresentative inputs*, such as using 1200 UTC observations for storms that will take place at 0000 UTC. The index may have a *weak foundation* with mathematical vulnerabilities and insufficient or biased verification metrics. Another problem is *noise* in multidimensional constructs such as the SWEAT index where a particular combination of input values or errors in the values may produce unrepresentative results.

During the 1960s and 1970s without the benefit of mature mesoscale forecasting techniques and desktop computer technology, index-based forecasting was in fact quite common. Techniques drew from dozens of stability and shear parameters, and in some respects they made up an ensemble method. However given the advances in conceptual models, nowcasting, and analysis technique, diagnostic parameters should be used as sparingly as possible.

The prudent forecaster must understand the workings of all diagnostic parameters that are part of the forecast, keep their use to a bare minimum, and treat them in the same manner as output from other analytical tools such as numerical models. While many of them, such as instability and shear measurements, are extremely useful a forecaster will also inspect the underlying fields.

3.3. ALGORITHMS. It is entirely possible to create a forecast based on a flowchart or a decision tree. This is essentially a diagnostic parameter whose output is entirely conditional on one or more factors. A number of noteworthy algorithms do exist, such as the 1979 Moller flowchart and the 1987 Colquhoun method. They are useful for illustrating important aspects of a decision making scheme but follow rigid paths that exclude dynamical and intuitive considerations. That said, algorithms are extremely useful when used for simple tasks, such as making an assessment of the hodograph. The potential for error rapidly increases when algorithms are used to take on more complex nonlinear work, such as predicting thunderstorm initiation or severe weather modes.

Meteorological cancer

A few weeks ago, I had the opportunity to visit an office on the east coast that recently had NEXRAD and other new systems installed. At mid morning there was a cloud mass over Ohio. When I asked a forecaster why it was there, the answer was, "it is residual debris from nocturnal thunderstorms associated with a dying front." Based on no echoes and a moth-eaten-looking cloud shield on the satellite picture, the forecaster said that clouds were moving slowly southeast and dissipating.

Looking at the 500-mb chart, it was apparent that these clouds were associated with a migratory temperature trough. With the help of diurnal heating afternoon thunderstorms were likely.

Several hours later driving west I encountered such heavy rain I considered pulling off the road until it let up.

Even with new tools available there is a need for in-depth diagnosis of more than mesoscale and local data. If forecasters aren't motivated to make good diagnoses there will be many in the new [model forecasting] era killing gnats and letting elephants run loose when preparing local forecasts.

LEN SNELLMAN, 1991
late meteorologist

Figure 1-5. Hand analysis. The author analyzing a severe weather situation in May 2003. As can be seen, attention to detail is more important than neatness or simply drawing lines for their own sake. *(Tim Vasquez)*

The forecaster's role

The qualitative diagnosis is exceptionally important for severe storm forecasting. In 2009, numerical models handle mesoscale and synoptic scale pressure, temperature, wind, and humidity fields fairly well, but continue to handle storm initiation and characteristics quite poorly. The human forecaster, using all available tools, is vastly more skilled at envisioning accurately how a storm situation will unfold and expressing it in terms of risk to life and property.

4. Subjective methods

A subjective forecast method applies comprehension, knowledge, visual-spatial thinking, and fluid reasoning. Since computers are not yet capable of these methods, the human remains a valuable part of the forecast process.. The forecaster applies a blend of analytic skills as well as intuitive skills, the latter being decisions which are largely unconscious and probabilistic in nature and which draw from experience.

Len Snellman in 1982 devised a useful framework for subjective analysis and diagnosis called the *forecast funnel*. This method assumes that an atmospheric process is only made possible by what's occurring at the larger scale, and that it makes sense to start with the "big picture" and work one's way down. This type of top-down technique suggests that the severe weather forecaster first start with global and synoptic scale charts and then work one's way down to mesoscale systems.

4.1. HAND ANALYSIS. Hand analysis of weather charts is popularly thought of as a process that simply enhances details on the chart for the benefit of coworkers and customers. However a skilled forecaster sees this differently. Hand analysis is an activity that heightens one's own situational awareness. It causes the analyst to immerse oneself in the data and integrate all of the information into a unified picture of the atmosphere. Information is gained by the process, not the result.

Proper symbology and neatness is not important except when illustrating for others. The chart should be used as a sketch board for assimilating data from all available systems, such as outlining outflow locations from radar loops onto the surface chart to provide some measure of continuity, and just jotting down notes. There is often very little need to create contours for upper air charts, because there is very little finescale observational data and forecast models have become quite proficient at estimating the best possible isopleths of height, temperature, moisture, and wind fields. However, locating jets, finding short wave troughs, and shading in important processes helps greatly with the mental diagnosis and should not be merely left to computers.

Overall, machine *analysis* of surface data should be avoided as much as possible, though *plotting* of raw data (i.e. adding raw data to the map, station by station) should always be left to computers since it is extremely labor intensive. The high data density offered in today's weather maps gives forecasters the opportunity to find a wealth of important patterns and phenomena taking place. When high quality data is present, the forecaster can do a better job than a computer at recognizing important features, just as an architect can do a better job at recognizing the sociological, aesthetic, and market value of a computer-aided house design even though the software program itself can easily churn out reams of engineering and carpentry data.

Soundings, radar data, profiler displays, and satellite pictures are also prime candidates for analysis and diagnosis. The sounding can be modified based on guesses and numerical data to obtain a forecast sounding for the time of maximum heating, and hypothetical parcels can be lifted. This is all easily done with a pencil and 30 seconds of time, and such an exercise is far more meaningful to an experienced forecaster than a number from a diagnostic index.

4.2. CONCEPTUAL MODELS. A conceptual model is a idealized model of a particular process that embodies its most representative traits. It serves as a framework for understanding a complicated system and predicting how it will behave. We use many of them in

Assumptions to remember

Though the basics of meteorology are beyond the scope of this book, here are some basic principles useful for understanding the material:

■ Pressure (p) decreases as altitude (z) increases. Likewise, pressure increases as altitude decreases.

■ Air that rises cools. Air that sinks warms. This is because rising or sinking moves the parcel into a region with different pressure, changing temperature due to the ideal gas law. We refer to this as adiabatic cooling and adiabatic warming.

■ Moisture in severe weather forecasting generally refers to the actual mass of water vapor in a parcel of air, expressed by dewpoint or better yet mixing ratio (the ratio of water vapor to air). Relative humidity is a different measure and relates more to the potential for air to produce cloud material.

■ Cooling when the relative humidity is 100%, i.e. saturated, usually causes gaseous water vapor to leave the parcel in the form of water or ice. The result is condensation into water or deposition into ice.

■ Condensation (gas to liquid), freezing (liquid to solid), and deposition (gas to solid) release latent heat. This adds warmth to the air.

■ Melting (solid to liquid), evaporation (liquid to solid), and sublimation (solid to gas) absorb latent heat. This cools the air.

this book to present complicated ideas in an understandable fashion. The well-known diagram of a supercell storm with all of its parts annotated is an example of a conceptual model.

While a conceptual model helps immensely with visualizing and memorizing important parts of a process, the danger is that it promotes "cookie cutter" forecasting where all processes must follow specific behaviors in order to be correctly predicted. Forecasters can avoid this by examining the constituent parts of a process and considering the meteorological significance of each contribution. This is known as *ingredient-based forecasting*. In other words, forecasters should strive to deal with the existing process rather than think of them in terms of conceptual models. However if the system is inherently too complicated or the role of the ingredients are not entirely clear to the forecaster, starting from a set of conceptual models may prove to be the wisest decision.

5. The balanced forecast

By far the most common forecast mistake is where forecasters omit the diagnosis due to hasty analysis, lack of understanding, or inability to synthesize the available information. In a form which has become rather endemic, the forecaster expresses strong skepticism and even open disdain for models — numerical and conceptual alike — but the forecast process clearly demonstrates great reliance upon them. This problem is often referred to as "meteorological cancer" as defined by meteorologist Len Snellman in 1977. It cannot be blamed wholly on the meteorologist. Many forecast offices are afflicted with technological and procedural design which hinders analysis, treats diagnosis as something the forecaster needs to be freed from, and emphasizes going straight to the numerical output to save time. An intense focus on modelling in the journals has also become a contributing factor during the past 30 years with case studies described in terms of arrays and diagnostic quantities rather than processes and mechanisms.

A meteorologist who has produced an poorly balanced forecast can accurately explain the details of what is taking place across Colorado and what changes the RUC is bringing, but is unable to accurately answer why surface winds are backing, or why temperatures are rising at Limon but not in Burlington. There has been a

---

The role of the human

For the foreseeable future, it is not likely that quantitative models alone will suffice for weather forecasting and so we as meteorologists and forecasters must continue to use qualitative information about atmospheric processes. Without the critical step of diagnosis, we are powerless to do so. Many advocates of new technology (who are not meteorologists and/or forecasters) do not understand what a vital link in the chain of scientific reasoning is the diagnostic step. Many see it as a boring, repetitive procedure ripe for automation. This would be tragic for the science of meteorology.

CHARLES DOSWELL III, 1986
The Role of Diagnosis in Weather Forecasting

Forecast discussions

I sometimes regret that with the coming of numerical weather prediction and a generally greater pressure on time in the forecast room, the daily map discussions have ceased. Although the forecasters are still taught Sutcliffe's thickness theory in training, I wonder whether they acquire the same insight into the way the atmosphere works that we gained from talking it over daily.

J.S. SAWYER, 1976
British Met Office forecaster

THE FORECAST PROCESS 13

thorough analysis and prognosis, but the diagnosis is clearly absent. The forecaster lacks insight.

In order to develop a grounded, solid forecast, the forecaster must achieve genuine quality and excellence in the three aforementioned areas: analysis, diagnosis, and prognosis. This excellence includes judicious use of diagnostics, models, and analyses. Soundings across the threat area must be analyzed. A thorough mental framework must be constructed as the analysis evolves. The RUC is used to estimate warm front motion but is set aside when storms are about to initiate. The palette of forecast tools and numerical guidance most be properly assembled, fully understood, and properly prioritized.

The final stages of diagnosis and prognosis are where the forecast process is shaped by increasingly important degrees of heuristic decisionmaking. This relies on two key factors: *expertise* and *experience*.

Expertise is measured by the *ability to draw upon and apply scientific knowledge*. Thus a meteorology degree alone does not equate to expertise but provides a foundation of skills and knowledge to allow for the development of expertise. The forecaster must keep abreast of journals, presentations, and conferences dedicated to forecasting and have an open mind for new ideas and concepts.

Figure 1-6. The technical library. An essential part of forecasting is keeping up on the latest knowledge and techniques. Fortunately the Internet has greatly facilitated this open exchange of information. In a sharp contrast to the policies embraced by many scientific and medical bodies, the American Meteorological Society has a very liberal policy of free Internet access to most scientific papers and conference preprints, with the exception of those appearing in refereed journals during the last 5 years. Even those can be purchased for $50/yr just to get the key journals: Weather and Forecasting and Monthly Weather Review. The National Weather Association's National Weather Digest is paper only for members, however the Electronic Journal is available at <nwas.org>. *(Tim Vasquez)*

> **Mesolows and tornadoes**
>
> Several investigators have pointed out that tornadoes and other severe thunderstorm phenomena are more closely related to small-scale meteorological features than to the larger scale of events usually depicted on the familiar synoptic charts. Means detected the presence of micro-lows or "tornado nests" which "frequently appeared an hour or two before the time at which tornadoes developed". Kraft and Connor describe "small lows" that occasionally "develop in the warm sector when conditions are favorable for major tornadoes". Whiting mentions the creation of "micro waves" along warm fronts in relation to severe weather. Each of these investigators identified severe convective phenomena with small-scale lows.
>
> BERNARD W. MAGOR, 1958
> SELS meteorologist

Experience, on the other hand, is not measured by gray hairs and the size of a job résumé. It's measured by the *ability to acquire insight from past decisions*. So there must be a long history of problem solving and the motivation to continuously re-examine consequences of past decisions. This requires an element of memory, clarity of mind, and passion. Storm chasers are particularly noteworthy for their sheer volume of meteorological experience because decisions are not forgotten at the end of a shift but tallied with great concern in terms of time, money, and personal interest.

Together, expertise and experience are the things that provide intuitive ability. Intuition is not simply an flash of insight or a lucky guess but rather a strong feeling based on firm but often unexplainable reasons. Intuition influences the decisionmaking process, leading to the best possible outcome. Heuristic decisionmaking is possibly where the right brain enters the forecast process and where the true art of forecasting emerges.

The proverbial meteorology graduates who can expertly solve partial derivatives for vertical motion but are "unable to forecast their way out of a paper bag" illustrate that the art of forecasting is not something that can be learned. Book knowledge and problem-solving ability are not the keys for success. The ability to apply these things, however, are. Meteorology must be observed, experienced, and studied in order to succeed in its understanding.

6. Scales of motion

Though most traditional forecasting work takes place at national scales, the nature of severe thunderstorm forecasting deals with many different processes with wildly varying lifespans and

Table 1-1. Orlanski system of atmospheric scales, with temporal approximations and some examples.

Common name	Term	Spatial	Temporal	Examples
Planetary scale	Macroscale-alpha	> 10,000 km	> 1 wk	Subtropical high, El Niño
Synoptic scale	Macroscale-beta	2000-10K km	> 3 d	Cold front, frontal low
Larger mesoscale	Mesoscale-alpha	200-2000 km	6 h - 3 d	Dryline, MCS, short wave
Smaller mesoscale	Mesoscale-beta	20-200 km	45 min - 6 h	Storm cluster, gust front, MCV
Storm scale	Mesoscale-gamma	2-20 km	5 - 45 min	Cumulus, thunderstorm, misoscale
Tornado scale	Microscale-alpha	200 m -2 km	30 s - 5 min	Tornado, microburst
-	Microscale-beta	20 m - 200 m	3 s - 30 s	Dust devil, suction vortices
-	Microscale-gamma	2 m - 20 m	< 3 s	Turbulence, roughness

Figure 1-7. The spectrum of scales. *(David Hoadley)*

sizes, from tornadoes and cumulus clouds all the way up to long waves which cover an entire ocean basin. To help categorize them for purposes of discussion and analysis, meteorologists use a system of atmospheric scales.

The system of atmospheric scale is calibrated in terms of spatial and temporal extent which describe its size and life cycle. For example, a small dust devil has a scale on the order of about 10 to 20 feet and a life cycle on the order of seconds. On the other hand, a frontal low has a scale of about 1000 miles and several days. The former is considered a microscale phenomena, while the latter is viewed as a synoptic-scale phenomena.

There is no real physical distinction between all of the different scales, which are essentially arbitrary classification schemes. The exception is that motion at the larger scales (planetary and synoptic scales) is strongly responsive to horizontal and vertical pressure gradients. It is in hydrostatic equilibrium and circulations are driven by baroclinic (temperature-contrast) instability. Motions that occur at smaller scales (mesoscale and microscale) are primarily responsive to things like buoyancy and orographic lift and are frequently non-hydrostatic.

The scale of motion generally accepted in modern forecasting is the Orlanski scale, published in 1975 (Table 1-1). A variation is the

A microscale perspective

One may divide the variables of the system into two classes, characteristic of the mesoscale and the microscale of the development. We shall refer to the mesoscale group all those properties of the bulk air which have to do with the hydrodynamical development on the scales of convective elements, that is, pressure, temperature, vertical velocity, and content of water in liquid and vapor state. To the microscale we refer the details of the droplet size distribution and the chemical properties of the aerosol, as well as any turbulence on scales that may interfere directly with the coalescence process.

CLAES ROOTH, 1959
Swedish meteorologist

An 1882 debate

Prof. Archibald said ... Mr. Symons had remarked that no one yet knew the precise distinction between a so called whirlwind and a cyclone or storm. It seemed to him that the ordinary nomenclature with regard to whirlwinds and storms required alteration; hurricanes, tornados, storms and tempests being spoken of as if they were synonymous terms. In a recent work by Prof. Ferrel of the United States Coast Survey he divided atmospheric disturbances into cyclones and tornados... According to Ferrel the smaller cyclones did not even begin [at a size] where the larger tornados ended, so that a great gulf [of scale] lay between them. There was also found to be an important qualitative difference between these phenomena since whereas the wind in the cyclone was a great deal affected by terrestrial rotation [viz. Coriolis effect] so that according to Ferrel's law if the air moved from one place to another it deviated to the right of its course in whatever direction it was moving. In the tornado owing to its small horizontal extent terrestrial rotation had no sensible effect on the motion of the air and there was thus a specific difference between the two.

J. WALLACE PEGGS, 1882 debating cause of damage in the October 1881 London gale

Fujita 1981 scale. It uses different ranges, adds two subdivisions, and progresses the second letter of "mesoscale" through the English-language vowels, ranging from macroscale at the largest size to musoscale at the smallest. This scale never gained widespread acceptance, but Fujita's *misoscale* classification has recently begun to see usage in the context of small circulations on the scale of single kilometers in size, detected with fine-scale radar systems. There have been proposals to use dynamic characteristics of weather systems to categorize scales, but so far geometrically-calibrated scales are considered to be most practical.

6.1. PLANETARY SCALE. The planetary scale encompasses systems that are around 10,000 km in size and have a life cycle of many weeks. The El Nino phenomenon is an example of a planetary scale system. Very little severe weather forecast work is involved at this scale, however planetary-scale systems do establish the basic underlying weather pattern affecting a region, such as whether the Great Plains is having a dry or a wet season.

6.2. SYNOPTIC SCALE. The synoptic scale covers a size of thousands of kilometers and a length of many days. Frontal systems, large polar air masses, and large hurricanes are examples of synoptic scale systems. It is rare that motion at the synoptic scale directly causes individual thunderstorms, however synoptic scale systems are highly influential in preparing the severe weather environment.

6.3. MESOSCALE. The mesoscale covers tens to hundreds of kilometers in size and a life cycle of hours. There is no rigid definition of what "mesoscale" is. The *AMS Glossary* defines mesoscale as ranging from "a few to hundreds" of kilometers in size. The two most popular scales, the Orlanski and Fujita scales, provide for a range of 2 to 2000 km and 4 to 400 km respectively. Consequently, the range of 10 to 1000 km is generally considered as "mesoscale" in severe weather forecasting. The term "subsynoptic" often appears in meteorology, but its only formal definition is in Fujita 1986 which specifies "disturbances too small to be depicted on synoptic charts", implying both mesoscale and microscale processes.

THE FORECAST PROCESS   17

6.4. STORM SCALE. The term "storm scale" refers to the scale of an individual thunderstorm. This covers the lower end of the mesoscale range (meso-gamma) and covers about 10 to 20 km in diameter.

6.5. TORNADO SCALE. When talking about wind flow, supercell structure, and radar interpretation, the term "tornado scale" does not refer to the Fujita damage scale but to the scales of motion associated with a tornado. This encompasses the micro-alpha scales and covers a size of single kilometers in diameter.

7. Wind expressions

This book contains numerous references to wind direction, shear, and vorticity, so it is extremely important that we stop and define these terms. It is strongly recommended that this section is studied carefully before proceeding further in the text.

7.1. WIND DIRECTION. It is important to remember that winds in meteorology are always referenced to the direction *from* which they blow. A wind barb is always drawn with the origin at the station

Hand analysis

The hand analysis process can be aided with objective analysis, which is a representation of a weather field using computer-based analysis. It can be argued that automated chart analysis saves time, however this overlooks the fact that forecasters differ widely in their ability to recognize important patterns in computer-drawn charts (Doswell 1986). In effect, this data does not enter the forecast process in a meaningful way unless it either feeds a numerical model or is scrutinized by the human forecaster and integrated into the mental subjective analysis. For this reason, many forecast offices refrain from producing machine isopleths and task the forecaster with "completing" the chart to mentally ingest the information.

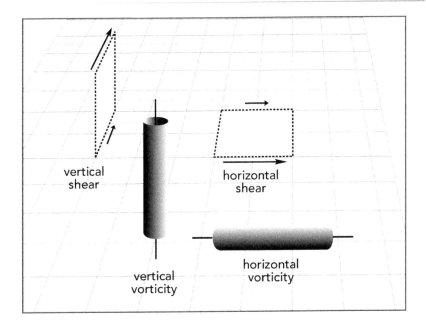

Figure 1-8. **Shear and vorticity.** This diagram illustrates horizontal and vertical shear and vorticity. Note that horizontal shear produces vertical vorticity. Likewise, vertical shear produces horizontal vorticity.

Figure 1-9. A shear vector (hollow arrow) shown as the difference between two observed winds (solid arrows) at different levels. Consider that this scale is calibrated in knots. This shows low-level winds out of the southwest (blowing northeastward) at 20 kt. Upper-level winds are from the northwest at 40 kt (blowing southeastward). Therefore the shear vector blows toward the south-southeast at 55 kt (i.e. the geometrical length, or magnitude, of this vector is 55 kt). Another way to think of this is with two hot air balloons, one at each level. Picture that the lower-level balloon has a rigid tower extending up to the upper level, and a device at the top of this tower connects to the upper-level balloon by a string. Initially the tower touches the upper-level balloon, but then we set the balloons free. The connecting string will orient itself in the direction shown by the hollow arrow here and the string will be reeled out at a speed of 55 kt.

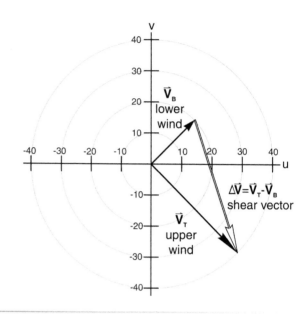

What it takes

A big factor in storm chaser success is a willingness to take a chance on your forecast. Consider 24 May 2008 as an example. A majority of chasers focused on Kansas and Nebraska based on model forecasts from the prior day. However, overnight convection drastically altered the distribution of moisture and instability, leaving the only undisturbed air mass in OK. The first supercells formed rather early in the afternoon, but provided quite the treat for those willing to revise their earlier expectations. Such chase successes go well beyond targeting a 24 hour forecast bullseye in a model, or simply driving to the western part of an SPC tornado probability max.

RICH THOMPSON, 2009
personal communication

location and the shaft drawn outward into the wind. The fletch containing the barbs is on the upwind portion of the shaft, so if the wind barb is likened to an archery arrow it would seem to fly downwind.

The exception to this rule is when describing wind as a vector, as is done when considering shear, hodographs, trajectories, and working directly with equations of motion. In this context we express the direction *toward* which the wind blows. Drawing from the origin to the endpoint, a wind barb points upwind, but a wind vector always points toward downwind.

The term downwind describes the direction toward which the wind blows, and upwind describes from where it originates. The term "left" and "right" is frequently used in respect to the wind, and this is always the direction which applies *with our back to the wind and facing downwind*.

Another quality of wind direction is *backing* and *veering*. Veering indicates a clockwise change in direction with increasing height, increasing time, or from point A to point B. Backing, on the other hand, is a counterclockwise change. The margin notes contain some important tips on using these expressions.

THE FORECAST PROCESS 19

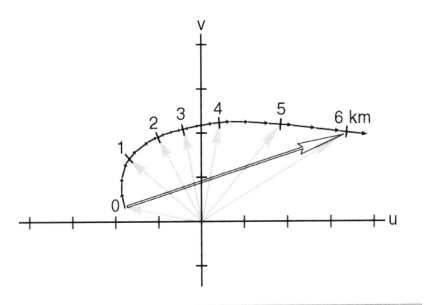

Figure 1-10. Hodograph essentials. Ground-relative observed winds are shown by the gray arrows; for instance, winds at 0 km are out of the east at 18 kt, and at 6 km are out of the southwest at 35 kt. The hodograph is shown by the black curved line, along with a multitude of hypothetical shear vectors which make up this line. Note how the individual shear vectors change direction, producing a curved line; this is the property of helicity. Bulk shear is shown by the hollow arrow and is simply the vector difference between the bottom and top of the layer without considering points in between. The bulk shear vector has a proportional relationship with storm cell longevity.

7.2. SHEAR. In meteorology, shear is simply a difference in wind (more accurately, a change in the wind vector) across a given distance. The term *vertical shear* refers to changes along a vertical axis, that is, between different heights, and is usually the form expressed in severe weather forecasting. The term *horizontal shear* refers to changes along a horizontal axis. Horizontal shear is not often used in severe weather forecasting but may describe sharp changes in wind across a front or outflow boundary. When the total shear is measured point by point through a layer (i.e. integrated) to show all changes in wind with height it is referred to as simply shear. The tool used to evaluate shear is the hodograph, which will be discussed in the Stability & Shear chapter.

7.3. SHEAR VECTOR. The simple difference between winds at two levels are expressed by a "shear vector". In practice, this is equivalent to the wind at the upper level as would be seen by an observer drifting in a hot air balloon at the lower level. Bulk shear is a common expression of a shear vector in forecasting, normally referring to the shear vector between 0 and 6 km AGL.

For example, consider a situation where winds near the ground are from the east at 40 kt and at the tropopause they are from the west at 40 kt. Simple vector math, which can be drawn on graph paper, shows that the shear vector points eastward at 80 kt. Thus

Small scale measurement

Weather maps often portray a uniform wind field across a forecast region, but step outside and the winds are ruffling one tree but not another. Likewise the atmosphere, especially on a severe weather day, contains relatively homogenous flow at the large scale but wind flow at scales of feet or mere miles that is often very irregular. This makes smaller-scale vorticity and shear quite difficult to accurately measure, not even considering the properties of parcels that are ingested by a storm. A sounding always contains some of this small-scale noise and can even be deceptive in its representation of parcels around a region.

The small scale

Without a radar array or a highly refined mesonet, it is impractical to sample all of the minor perturbations in the wind flow across a forecast area down to the wind gusts and dust devils that mark the smallest of these features. This is unfortunate because stretching of very small-scale vorticity maxima are associated with the development of rotation within some thunderstorms as well as a considerable number of landspout and waterspout reports.

an observer drifting in a hot air balloon just above the ground would see the cirrus clouds moving from west to east at a velocity of 80 kt. Since storms care only about storm-relative winds and ground-relative frames of reference are meaningless for them, this example provides a first glimpse at learning to "think like a storm"! Hodographs allow us to evaluate wind in storm-relative frames of motion.

We can also express directions relative to the shear vector. The term *downshear* represents the direction of the shear vector, i.e. the direction in which it points. The term *upshear* refers to the opposite direction. Much like "left" and "right" in terms of wind flow, "left" and "right" are treated as if we are standing with our back to the wind, standing on the shear vector origin and looking toward the head, to where the wind is blowing. In the example above, the cirrus clouds always move away in the *downshear* direction regardless of which way the winds are blowing.

7.4. SPEED SHEAR. When shear vectors change in magnitude with height (remember here we are considering multiple layers) but are all oriented in the same direction, this describes a straight line hodograph. It produces an environment where speed shear is the dominant process. This can occur when the wind direction is the same at all heights, a condition known as a *unidirectional* wind field. But a unidirectional wind field is not a requirement for a straight line hodograph! Winds relative to the ground might appear to change direction and speed with height but when considered from a storm-relative frame the winds might be observed to all be blowing in the same direction.

7.5. DIRECTIONAL SHEAR. When shear vectors show directional changes with height, this reflects a property known as "helicity". The change in shear vectors produces curved segments on the hodograph. The greater of a geometric concavity that they produce, the greater the helicity in that layer. This property has a unique role in forecasting. It will be discussed in great detail in Chapter 7, Stability and Shear. The important thing to remember is that curved hodograph line segments are indicative of a layer with helicity. In the northern hemisphere, clockwise turning of a hodograph segment with height corresponds to positive helicity. This is loosely equivalent to veering of wind with height (i.e. backing when moving from higher to lower levels).

7.6. VORTICITY. Another expression of shear is vorticity, which is a measure of the tendency for air to spin. It can be thought of as the spin which would result if a coffee can of indefinite size is placed in the flow and its round side is free to rotate according to the wind it encounters. While simple shear is often directly responsible for vorticity and can often be a good proxy for vorticity, curvature in the airflow and even can also contribute to spin. Thus, vorticity is the sum of shear and curvature.

Vorticity is expressed in revolutions per second, but since revolutions are dimensionless, vorticity is expressed as seconds, s^{-1}, also known as simply "units". A tornado, for example, has a vorticity of about $1 s^{-1}$, or one revolution every second. Synoptic scale features are by convention measured in units of $10^{-5} s^{-1}$, which is about one revolution per day.

Vorticity considered along a vertical axis, like a spinning toy top, is referred to as *vertical vorticity*. Vorticity measured on a horizontal axis, like a rolling log, is referred to as *horizontal vorticity*. Vorticity about an axis parallel to the wind flow is called *streamwise vorticity*. This can be likened to a paddle wheel that is broadside to the wind. Vorticity can even be measured on a horizontal axis perpendicular to the wind flow, yielding *crosswise vorticity*. Streamwise vorticity is indicative of directional shear with height, while crosswise vorticity indicates speed shear with height and no directional shear (see Figure 1-11).

There are also two types of vorticity expressions. The pure measure of vorticity, considering only shear and curvature, is known as *relative vorticity*. This is used in most storm-scale analysis. When the earth's rotation is added in, shear, curvature, and planetary vorticity are combined to obtain *absolute vorticity*. This is widely used in synoptic-scale analysis.

It must be mentioned that vorticity carries a double meaning in forecasting. Since the 1950s it has provided a popular diagnostic tool for vertical motion since the mid-tropospheric vertical vorticity bears a correlation to the amount of subsidence or lift in the atmosphere. This technique has created terms such as "positive vor-

---

Cutting through confusion

Backing and veering can be expressed as a function of either distance or time. This is sometimes a source of confusion for beginning forecasters, since "backed winds" are often spoken in the same breath as supercells.

Consider a situation where the winds aloft blow out of the southwest and are held constant. But the surface winds at Omaha are blowing out of the southeast, then become easterly. This indeed is more conducive to severe weather. This change is a counterclockwise shift with time. So winds have indeed backed with time.

Analysts reviewing surface maps might compare other stations in the area and note a counterclockwise shift when moving a finger from Topeka to Omaha or from Grand Island to Omaha. This shows backing with distance, e.g. backing as we approach Omaha.

Now when we consider the vertical wind profile, it shows a clockwise change with height, thus it is a veering profile. But the backing of winds at the surface with time actually creates a more sharply veered profile with height! Veering with height favors cyclonic storms, in spite of the connotation that accompanies the term "backing".

This explains why it is often a good idea to be clear when using "backing" and "veering" and indicate whether you're referring to a change with height, a change over time, or a change between two points.

ticity advection", PVA, and "vort max" and will be described in the forecasting chapter.

However in storm forecasting, the actual properties of vorticity, particularly at small scales and within the inflow layer, are considered important when considering small-scale circulations, hodographs, inflow, and tornadogenesis. Streamwise vorticity is considered especially important in storm formation. It will be described in the shear and stability chapter.

The sign of vorticity, in other words whether it is positive or negative, is indicated by the right-hand rule. This rule states that if the fingers are curled and used to represent air flow, with air moving from the base to the tip of the finger, the vorticity will be positive if the thumb is facing up and negative if facing down. For horizontal vorticity, a thumb facing into the view or into the page indicates positive vorticity. ¤

THE FORECAST PROCESS 23

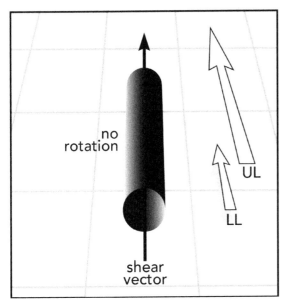

Figure 1-11a. Streamwise vorticity in pure speed shear. Since no motion occurs orthogonal to the lower-level (LL) and upper-level (UL) winds, the tube does not rotate.

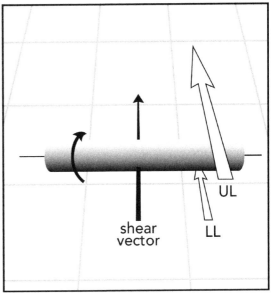

Figure 1-11b. Crosswise vorticity in pure speed shear. The differential motion of the lower-level (LL) and upper-level (UL) winds causes crosswise vorticity to develop..

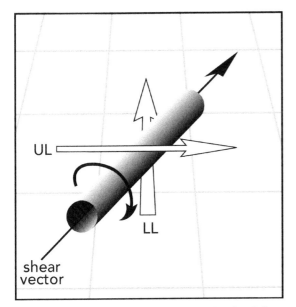

Figure 1-11c Streamwise vorticity in pure directional shear. Differential motion between low-level (LL) and upper-level (UL) winds causes the tube to rotate.

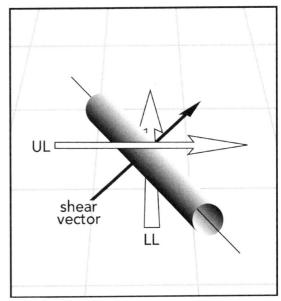

Figure 1-11d. Crosswise vorticity in pure directional shear. Low-level (LL) and upper-level (UL) winds oppose each other in rotating the tube, therefore it does not rotate.

2

THE THUNDERSTORM

It is also [August], the beginning of the thunderstorm season, and here in the region between the eastern seacoast and Cleveland the storms are surpassed in ferocity only by those truly evil monsters whose lair is farther to the south, between Washington and Memphis. Others of like pugnacity lurk about in the midwestern states, but a thunderstorm is a thunderstorm no matter where encountered, and all of them have the disposition of a Caligula . . . Here the route [to Newark] follows the Hudson River southward, passing over the Catskill Mountains, where the legendary ghosts of Dutchmen bowl with such abandon. In this region a special combination of terrain, summer heat, and humidity serves to produce spectacular monuments of clouds. Pilots who would much rather prefer admiring this grandeur from afar frequently refer to such formations as sons-of-bitches. The pilots know they are not to be trifled with.

— Ernest K. Gann, 1940s airline pilot
Fate Is The Hunter, 1961

To a farmer, the thunderstorm is a big black cloud full of rain. But to a meteorologist, a thunderstorm embodies a multitude of processes that occur when the atmosphere is unstable enough to allow saturated air to rise rapidly. Rain, hail, and lightning are just the end result.

Storms are the result of convection: the upward transport of heat by buoyant lift. The fundamental building block of a thunderstorm is a *convection cell*. This is the circulation caused by a convective updraft and its resulting downdraft. A thunderstorm can contain one cell, though in reality it is often made up of multiple cells, thus the name "multicell".

1. Storm morphology

The 1946 Thunderstorm Project accurately determined the existence of the updraft and downdraft in a thunderstorm, and researchers Horace Byers and Roscoe Braham identified the stages of a storm: cumulus, mature, and dissipating. These concepts are still in use today to form the building block of an updraft-downdraft life cycle.

1.1. CUMULUS STAGE. When solar heating is introduced, the Earth's surface warms, conducting heat into the contact layer with the atmosphere. This rises upward in the form of thermals. If sufficient moisture is present, cumulus clouds form. If they continue growing, the cumulus will grow vertically, producing towering cumulus that occupies a deep layer in the troposphere. Precipitation particles are forming within the cloud but on radar only show weak reflectivity at about 3 to 5 km above the surface.

However the cumulus continues growing and begins transitioning into the mature stage. Cloud droplets collide and coalesce at the top of the cumuliform towers, forming rain droplets. Ice particles may also be produced. All of these precipitation particles grow and multiply in number. As they grow from microscopic to rain-drop sized particles, their fall speed increases from a negligible speed to one that is quite significant. The fall of these particles, through friction, causes drag on the surrounding air. Partial evaporation chills the surrounding air and adds to its density, accelerating the fall of this mass. This is the production of a downdraft.

An 1830 explanation

Before the thunderstorm is announced, we shall perceive in the distant perspective horizontal lines, parallel to each other — a kind of dense vapour plane; this plane becomes still more dark and dense, arches of clouds rest upon it, the upper parts even and well defined, but dark and threatening at their base. Sometimes this aerial volcano seems to swell in magnitude by attracting other clouds in its vicinity, which appear to fall into it, while at other times it increases in magnitude without any such visible cause.

At length the cloud is put in motion; the thunders roll and the lightnings flash, and the rains descend, till, having spent its materials, the nimbus separates into fragments, which melt away in the aerial abyss, and discover to us a serene sky.

JOHN MURRAY, 1830
A Treatise on Atmospherical Electricity

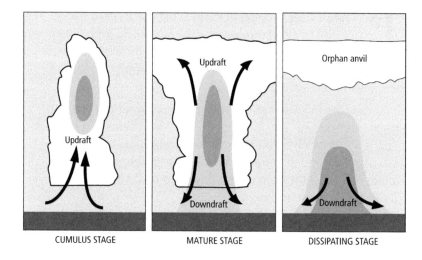

Figure 2-1. **Simple life cycle of an ordinary thunderstorm**, essentially the model proposed by the 1940s Thunderstorm Project and which is still widely considered to be the fundamental building block of a thunderstorm's life cycle. Long-lived thunderstorms are those that sustain some form of a mature stage indefinitely *(Tim Vasquez)*

1.2. MATURE STAGE. A thunderstorm soon reaches mature stage, containing a significant downdraft and a significant updraft. Precipitation or virga is usually widespread and the storm is easily detectable by radar. It is also at this point where the storm has the best potential to produce tornadoes or very large hail.

A large mass of falling precipitation and rain-cooled air, however, is not compatible with a rising updraft. The updraft finds itself increasingly filled with rain and cool air, and it becomes less buoyant and eventually ceases to exist.

1.3. DISSIPATING STAGE. The storm cell has entered its dissipating stage when the storm consists almost entirely of a downdraft. The heaviest surface wind and precipitation, ironically, occurs at the beginning of the dissipation stage since at this time the precipitation cascade is still above the ground. The arrival of the core at the ground is sometimes referred to as a "collapse". The downdraft gradually diminishes in intensity until there is no longer precipitation in the downdraft. The radar continues to detect the storm until all precipitation particles are gone. This can take hours to occur. Still, the decaying remains of the storm may persist for days, with extensive convective debris fields of altocumulus and thick cirrus.

2. The updraft

The updraft is produced by a buoyant, rising parcel of air moving upward into the free atmosphere. When the parcel rises, it cools adiabatically. If the parcel contains moisture, then at some point it will cool to its dewpoint temperature, at which point it saturates and begins producing condensation of droplets or sublimation of ice into the parcel itself. This is seen as the creation of cloud particles. In a weak updraft these particles all together will form a single cumulus or stratocumulus cloud, but in a strong updraft it will form a cumulonimbus cloud. It should be noted that the rising thermal or updraft does have some interaction with the air around it and will ingest some of it around its periphery as it rises, a process known as *entrainment*.

If the updraft is strong and has been rising through the troposphere, it soon reaches its equilibrium level, where it encounters relatively warmer air in or near the stratosphere and is no longer cooler than its environment. It does continue to rise, forming a towering cumulus cloud, which develops precipitation particles and becomes a cumulonimbus with hard edges, a cloud species known as *calvus*. As ice crystals rapidly form at the top of the cloud it produces fibrous edges, and at this time the cumulonimbus may be described by the species name *capillatus*. The tower continues rising to near the tropopause, where the rising parcel suddenly becomes colder than the environment. It loses momentum, forming

Weight of a thunderstorm

An average summertime thunderstorm cell has a mass of about 10 billion kilograms. This on the same order as that of the Great Pyramid of Giza in Egypt and about ten times that of the Sears Tower in Chicago.

Figure 2-2. **Large barrel-shaped updraft** on a classic supercell near Anadarko, Oklahoma, looking southwest. Note the large wall cloud underneath and the tail cloud leading off to the right side of the image. This storm formed on 3 May 1999 during the Oklahoma superoutbreak. *(Tim Vasquez)*

an overshooting top at its highest prominence. It then sinks back down underneath the equilibrium level and spreads away from the storm to form an anvil cloud. This spreading away of an anvil is caused by accumulation of mass at the top of the cloud due to the updraft. The cumulonimbus may then be known by the species name *incus*. In an atmosphere with calm wind, anvils can be forced to diverge horizontally for tens of miles, and if carried by the upper-level wind, can also spread downwind for hundreds of miles.

The bottom of the cumulonimbus cloud is manifested by an extensive area of dark cloud base known as the rain-free base, a term which references its relative lack of rain compared to other parts of the storm. It is almost always the darkest part of the cloud and may appear quite ominous. It is at the downdraft interface along this rain-free base where the greatest risk of tornadogenesis is likely.

Young but very tall cumulus towers may be adjacent to the updraft area, forming a cloud field known as the *flanking line*. Also known as a feeder line, this flanking line is typically on the equatorward side of the storm given typical moderate to strong wind profiles aloft. Flanking lines which are made up of discrete, well-spaced towers will result in a pulsed life cycle and a lower probability of severe weather. Flanking lines which consist of towers very close together or merged will result in more of a steady-state updraft and are often associated with supercells.

Figure 2-3 (lower left): Ragged rain-free base on the tail end of a severe squall line, 5 April 1988 in Greenville TX, looking northeast. A wall cloud can be seen in the distance, and an intense downdraft and precipitation area further behind. *(Tim Vasquez)*

Figure 2-4 (lower right). Laminar rain-free base on a strong multicell storm near Logan NM, 29 May 1997. *(Tim Vasquez)*

THE THUNDERSTORM   31

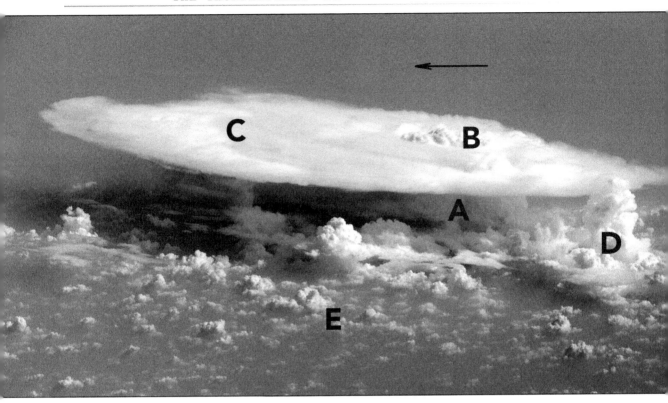

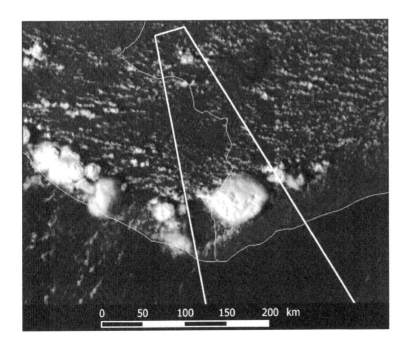

Figure 2-5. **Storms rain down on Guerebape, Ivory Coast,** a major palm oil production region in Africa. Observed on 5 Feb 2008 1510 UTC from the International Space Station (ISS), looking south. Corresponding visible satellite imagery is shown on left. A line of updraft towers (A) produces weak overshooting tops (B). The moisture from the updraft spreads in the form of an anvil cloud (C) primarily in the direction of the upper-level winds (arrow) and precipitation falls (downdraft areas are blocked by the anvil). An incipient cumulonimbus cloud is at (D). Low-level tropical moisture and haze (E) with embedded streets of cumulus contrast greatly with the drier, transparent middle and upper troposphere. NASA photo records list this shot as over Senegal but that is actually the approximate ISS nadir. *(NASA/ISAL)*

Figure 2-6. **Well-developed flanking line** on a multicell storm east of Hominy OK on 22 April 1999. This view looks southeast on the back side, showing towers rising to nearly 30,000 ft MSL. Note the discrete spacing of the towers, which suggests more of a pulsed convective mode. *(Tim Vasquez)*

The cumulonimbus tower may show *helical striations* on the side of the cloud which give it a corkscrewed appearance. If this is present it is a visual manifestation of the rotating updraft and indicates the presence of a mesocyclone.

3. Mesocyclone

"Tornado cyclones" embedded in the storm, much larger than a tornado but smaller in scale than the storm itself, have been noted on barographs since the late 19th century. By the 1960s, this type of small-scale cyclone was given the name *mesocyclone*. The actual circulation was first measured directly in 1971 using Doppler radar and was observed to be particularly strong in the mid-levels of the storm. A large body of research during the 1970s established the framework of the mesocyclone and tied it to the rotating updraft. In the mid-1970s the mesocyclone was established as a defining characteristic of the supercell model.

The mesocyclone normally is strongest in the mid-levels of the storm where tilting is at its maximum, at an altitude of a few kilometers, and has a size of several kilometers in diameter. It has a life cycle of about 30 minutes or more. The mesocyclone is detectable with automated radar algorithms or by direct observation of a velocity couplet. A mesocyclone should not be confused with a "mesolow", which occurs outside of thunderstorms and is on the order of about 100-200 km in diameter.

The classic supercell mesocyclone forms when streamwise vorticity in the fair weather environment preceding the storm is ingested by the storm updraft, tilting the flow axis from horizontal to vertical. The horizontal vorticity thus becomes oriented along a vertical axis. Stretching of this vortex, which occurs when one end moves in an opposite direction from the other (i.e. one end is being accelerated by the inflow and updraft) increases the angular momentum of the vortex. The mesocyclone and updraft are further enhanced by other processes within the storm.

The mesocyclone is the breeding ground for supercellular tornadoes. These typically form just upshear of the mesocyclone where the rear-flank downdraft and inflow air interact. Recent studies have found that about 26% of all mesocyclones are associated with tornadoes. A well-established mesocyclone, however, is important. If it appears on radar for about 20 minutes or more, it is more likely to be associated with the development of tornadoes.

4. Downdraft

A downdraft is initiated by the assortment of precipitation particles created by the updraft: supercooled water, ice crystals, graupel, and hail. The first particles to form are cloud droplets, which measure about a hundredth of a millimeter in size. They have a very weak fall speed which is overwhelmed by the updraft, so they tend to rise with the updraft. However as collision, coalescence, and other precipitation growth mechanisms take hold, these cloud droplets rapidly grow and either fall within the updraft or outside of it. By colliding with air molecules as they fall, they drag the air down with them. Additional downward motion is produced if the downdraft entrains dry air, because some of the precipitation evaporates and increase the density of the air mass, increasing its negative buoyancy.

Rain-free base came into common usage by 1978 in spotter training. The concept is generally credited to John Marwitz, August Auer, and Donald Veal in 1972, who described a large, flat cloud base that corresponded to the updraft.

Flanking line was coined by former SELS forecaster Ferdinand Bates in 1967.

What's in the updraft?

The updraft core always contains a certain amount of condensed precipitation. This tends to "weigh" the updraft down and reduce buoyancy of the rising parcels. The updraft is also affected by entrainment of relatively cool air that surrounds the storm.

The updraft typically contains large amounts of supercooled water droplets. Very large zones of supercooled water provide highly potent ingredients for hail growth (see Chapter 5, Hail). Though the weak echo region does not initially contain hail, hailstone embryos do cross the tops of this zone and grow rapidly through accretion.

Downburst mechanisms

There are several mechanisms that can affect downdraft strength. In cases where dry air entrainment and precipitation evaporation are equally important (as in a case with dry mid-level air, and relatively dry low-level air yielding a high LCL), the situation is referred to as a wet-dry hybrid.

Not entirely a tall tale

Shortly after midnight on the morning of June 15, 1960, under clear skies and otherwise normal conditions, a damaging, scorching northwest wind struck terror and near disaster to a 25-mile stretch across the northwest side of Lake Whitney for nearly 3 hours. It was like any other Texas night in mid-June. The temperature was in the 70s, the stars were out and a light breeze was blowing. There had been some lightning earlier, but no one paid much attention to it. Then without warning... it struck. A searching blowtorch-like wind hit with speeds estimated at 80 to 100 mph, and the temperature jumped from near 70 to 140 degrees! The Mooney Village Store lost the roof and was badly damaged. . . Fire sprinkler systems were set off, car radiators boiled over, and panic-stricken women were crying, thinking the end of the world had come.

HAROLD TAFT & RON GODBEY
Texas Weather, 1975

A storm can be a prolific producer of downdrafts. Furthermore if the rate of outflow production substantially exceeds the rate of inflow, the storm is said to be outflow dominant. Storms that are outflow dominant tend to quickly undercut updrafts and if the updraft does not propagate or form elsewhere near the storm, the storm dissipates.

4.1. OUTFLOW. When the downdraft reaches the ground, it spreads horizontally, forming a pool of air known as outflow. The leading edge of the downdraft is referred to as an outflow boundary, marked by a gust front. The surface of the outflow is considered to be a zone of negative horizontal vorticity. It can be a source of vorticity for tornadoes.

4.2. HEAT BURSTS. Early in the 20th century, local legends existed of heat bursts that destroyed crops and broke mercury thermometers. Understandably, these reports were met with skepticism, but on 4 May 1961 the Oklahoma mesonet made the first instrumented detection of a heat burst event. Since then, heat bursts have become a regularly observed phenomena and an actual forecast issue.

The classic heat burst occurs late at night with a dissipating storm overhead. A quiet night with thunder and lightning rumbling in the sky gives way to gusty winds, a sharp increase in temperature, and a sharp decrease in dewpoint temperature.

The heat burst is explained by a downdraft which descends and reaches a stable nocturnal inversion near the ground. The air within this inversion is forced to subside, and it warms at the dry adiabatic rate. As a result, the observed effect of the heat burst is a temporary elimination of the radiational inversion, bringing surface temperatures briefly to what they were during the previous morning or afternoon. The nocturnal inversion quickly rebuilds over the next couple of hours and temperatures fall once again.

Wake lows that follow MCS systems have also been correlated with heat bursts, but are probably more rare. This type of heat burst occurs due to coupling of the downward rear-inflow jet with downdraft activity at mid- or upper-levels.

Figure 2-7. **Decaying cumulonimbus cloud** forming an orphan anvil. Photographed in the Ivory Coast, Africa on 13 April 2008. *(Paul Morley)*

4.3. DOWNBURSTS. Exceptionally strong downdrafts are given the term *downburst* (Fujita and Byers 1977). They are responsible for the vast majority of damaging straight-line wind events around thunderstorms.

When a storm system produces a family of downbursts with extensive damage, the event is referred to as a *derecho*. These are covered separately in this book. On the other hand, a single concentrated downburst of very small scale, usually less than 5 km or less, is referred to as a *microburst*. Since the 1970s, microbursts have captured news headlines after causing a number of airliner crashes. In a microburst, significant wind velocity changes can occur across an even smaller distance than in a conventional downburst. This causes aircraft to lose airspeed so rapidly that the plane stalls before the flight crew has a chance to react.

The basic set of conditions favorable for downburst and strong outflow are largely centered on strong evaporational cooling of the downdraft and precipitation loading.

36 THE THUNDERSTORM

If an updraft rises into dry mid-level air, it entrains this air into the updraft. This dry air then mixes with precipitation particles, which then evaporate or sublime and remove sensible heat from the air. This increases its density and downward momentum. Conditions favoring entrainment-augmented downbursts are *significant low-level moisture* but *very dry mid- and upper-level moisture*, a common configuration on the Great Plains.

Another source of evaporational cooling is when significant precipitation actually falls into a dry layer. The evaporation in this layer accelerates the downdraft and precipitation cascade. This type of situation occurs with elevated moisture, such as in mountainous or desert regions, including the so-called "inverted-V" sounding profile. This type of downburst is specifically called a *dry downburst*.

Sublimation actually generates a stronger cooling effect than evaporation does. Therefore if a large proportion of downdraft material consists of hail or snowflakes, the downdraft will be enhanced.

Figure 2-8. Damaging derecho event on 15 June 1995 in upstate New York. This radar frame is for 5:33 am EDT. Seven hours previously, this line was forming over Lake Superior. The track of highest winds extended east-southeast from Lake Ontario (top left), destroying over 1500 square miles of forest. Seven people were killed and the damage tally, partly from lost timber, exceeded half a billion dollars. Reflectivity shows nothing particularly unusual, but the forward speed averaged 70 mph. At the time of this frame, high winds were affecting a broad swath along the Interstate 90 corridor.

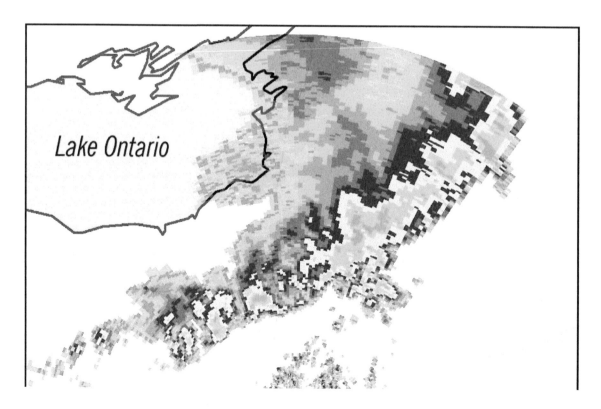

The greater the water and hailstone content in the downdraft, the more precipitation loading occurs. The falling particles impart momentum to the air they fall through, forcing the air to move downward too. Precipitation loading is boosted by whatever puts the greatest mass of precipitable water into the middle and upper portions of the cumulonimbus. Very moist inflow with high dewpoints yields the greatest chance for high precipitable water, but if this moisture extends to a considerable height then it can interfere with entrainment of dry air. Likewise if a downburst originates at high altitudes, such as when a strong updraft lofts precipitation particles to higher altitudes than usual, the greater fall distance can allow the downburst to build more momentum than would otherwise be expected.

There are also special mesoscale processes which enhance the downburst and the resulting outflow. These are described in the chapter on MCS storms.

Backing and veering

Never confuse backing/veering with height with backing/veering at the surface or backing/veering with time. It is important to understand which coordinate system is being used before interpreting the meaning. When surface plots show locally-backed winds or winds that become more backed with time, then on the hodograph we will see this as enhanced veering with height between the surface and levels aloft.

5. Storm motion

Storm motion generally follows the mean winds in the layer which contains the cumulonimbus cloud. Also it must be remembered that the storm cloud may be made up of many updraft-downdraft pairs, and as a result the storm also has a tendency to shift to wherever the newest, strongest cells are developing. This is referred to as *propagation*. Consequently, storm motion is the sum of the both the storm-layer wind vector and the propagation vector.

As early as the 1940s, American meteorologists discovered that tornadic storms often moved to the right of the mean environmental winds (i.e. toward the east or southeast rather than the northeast). This characteristic came to be known as *deviant motion*. In the 1950s and 1960s, theories were put forth to account for the

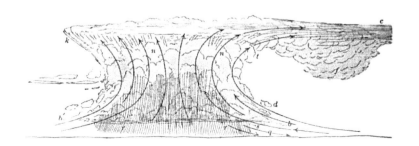

Figure 2-9. In the 19th century the general structure of thunderstorms was fairly well understood. This depiction was created by Harvard scientist William Davis in 1894, building upon a similar model by German researcher Max Moller in 1884. Davis observed that "some of these towering cumulus reach a greater size than the rest ... then the cloud assumes the familiar anvil form, so commonly associated with distant thunderstorms." *(Davis)*

propagation vector. Some treated the updraft as an obstacle in the environmental flow and applied principles like the Magnus effect which would enhance propagation of the storm in a particular direction. The Magnus theory proposed that since the right side of a cyclonically rotating updraft moved in concert with the environmental winds that it induced low pressure on the right flank and high pressure on the left flank, resulting in favored propagation to the right. Though this idea is attractive in its simplicity, it was largely discounted after the 1970s.

Since the 1980s it has become clear that deviant motion in supercells, that is, in storms with rotating updrafts, is the result of a specific configuration of vertical pressure gradients within the storm. Klemp and Wilhelmson's studies in the 1970s and 1980s used numerical modelling to demonstrate that veering winds with height* (i.e. southeasterly becoming southwesterly aloft) consistently favored production and intensification of deviant right $_{(NH)}$ movers, while backing winds with height* favored production and intensification of deviant left $_{(NH)}$ movers. With no change in wind direction with height, neither storm was favored and the storms had tendencies to split into deviant left-moving and right-moving pairs, each a mirror image of each other. Supercell movement along the propagation vector is usually continuous with respect to time.

Linear updraft theory shows that pressure perturbations in the storm are proportional to change in shear vectors with height, augmented by updraft strength, and occur parallel with the shear vector. Furthermore a change in direction of the shear vectors with height redistributes these pressure perturbations in the vertical so that instead of the same perturbations stacking vertically with height, their locations change with height, amplifying vertical motions that would not otherwise be there.

Even if a thunderstorm does not acquire rotation, its propagation vector can be greatly influenced by where the precipitation cascade falls relative to the storm, the behavior of the storm's outflow boundary, and the characteristics of the storm structure itself. Propagation of non-supercells along a propagation vector is discrete, that is, it occurs in intervals, usually with a cycle of about 10 minutes as each new cell forms, though squall lines can often adopt continuous propagation modes.

On storm propagation

I believe there is universal agreement that a clockwise-curving hodograph favors right-moving storms. Furthermore there is universal agreement that enhanced vertical pressure gradients on the right flank of on updraft existing in an environment characterized by a clockwise-curving hodograph produce new updrafts to the right of old updrafts, and thus, rightward propagation.

The only questions that remain are, in my view, technical ones concerning the relative roles of certain terms in the pressure equation that account for the rightward bias in vertical pressure gradient (see Weisman and Rotunno 2000 J. Atmos. Sci.)

RICHARD ROTUNNO, 2009
personal communication

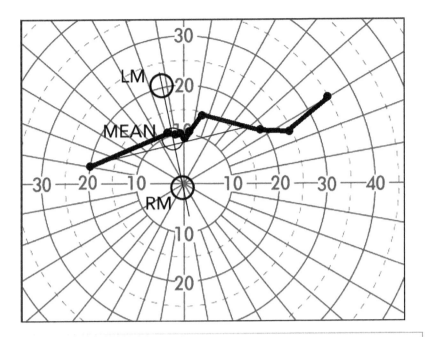

Figure 2-10. Modified hodograph for the afternoon of 1 August 1985 at Cheyenne, Wyoming. Note how the right motion vector results in an almost stationary ground-relative movement. The old 30R75 method would have yielded a ground-relative movement of at least seven knots. Storms developed over Cheyenne and became severe, producing hail that lasted for hours and accumulated in large drifts. At least 12 were killed and $43 million in damage occurred. This hodograph also shows a straight-line characteristic, which favors splitting storms.

Though propagation vectors are clearly influenced by internal processes within the storm, they can also be influenced by external processes. The storm may propagate preferentially along a pre-existing boundary, which greatly favors the development of new updrafts. Changes in the storm environment may also influence the behavior of internal processes within the storm, affecting its movement.

6. Splitting storms

Splitting is the division of a storm into two separate storms, one left-moving and the other right-moving. Each storm is structurally a mirror image of the other one. In the northern hemisphere, the left-mover contains anticyclonic storm-scale winds and moves to the left of the mean tropospheric winds (i.e. usually northward, while the right-mover has cyclonic storm-scale winds and moves to the right (i.e. usually eastward or southeastward). In the majority of such northern hemisphere severe storm cases where a split occurs, the right split is severe or tornadic while the left split contains only hail or quickly weakens. But in a fraction of such cases, the left split produces a long track of significant hail while the

Ekman spiral

The Ekman spiral is a natural tendency for the winds to veer with height in the northern hemisphere through the boundary layer (usually the lowest couple of km). This is because friction disrupts geostrophic balance close to the ground, slowing air parcels, diminishing Coriolis force, and allowing air to flow more directly toward low pressure.

If we descend from the free atmosphere where winds are moving west-to-east, we might find near the ground that winds prefer to flow southwest-to-northeast instead of west-to-east. This appears as locally backed winds near the ground which veer with height. This natural veering tendency greatly favors right-moving storms in the northern hemisphere.

right split weakens. In either case, the left split rarely produces any tornadoes.

Splitting modes are associated with straight-line hodographs, which can result from unidirectional winds but can also arise from weak mid-level flow and other factors that cause a fairly straight hodograph plot. If the environment contains speed shear, horizontal vorticity is present. Increasing winds with height will cause this vorticity to rotate with the top of the "vorticity tube" moving downwind and the base moving upwind. When the updraft builds, the horizontal vorticity axis is stretched into the vertical, forming a vortex on each side (looking downwind) of the updraft: anticyclonic on the left$_{(NH)}$ side and cyclonic on the right$_{(NH)}$ side.

Originally it was thought that the descending precipitation cascade or a rear-flank downdraft "broke" the vorticity tube into two segments, causing a storm split, but splitting has actually been observed in numerically simulated storms without any precipitation. It has been shown however that nonlinear pressure perturbations cause low pressure to develop at each of the vertical vorticity cen-

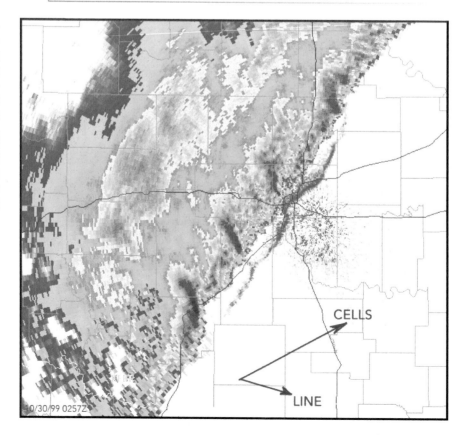

Figure 2-11. Serial derecho moving through central Oklahoma on the evening of 29 October 1999. Note how the line is broken up into a series of bow echoes. Widespread wind damage was reported through the region. The system also exhibits a pronounced trailing stratiform region.

ters. This causes the storm to propagate toward both of these areas, and this process causes the split.

In any case, the split occurs very early in the storm's life cycle. In the northern hemisphere, the right-moving split is more likely to produce severe weather; the left moving split moves faster than the right mover and its severe weather is usually confined to hail, though anticyclonic tornadoes can occur. Both the left and right mover are mirror images of one another. Their precipitation areas helps propagate the cells in their respective directions.

Splitting tends to result in a diversity of storm motions and greatly increases the chance of cell collisions and seeding of other cells. This may augment the precipitation efficiency of seeded cells and may have implication in hail forecasting. It also complicates the nature of spotter positioning, as spotter coordinators either need to recognize and emphasize the right$_{(NH)}$ splits or on left $_{(NH)}$ splits position chasers on the northern $_{(NH)}$ flanks of the storm to correctly identify key updraft areas.

Classification schemes
The air mass-frontal-squall line archetypes were in widespread use from the 1940s through the 1960s. This classified storms based on the prevailing weather pattern rather than storm structure. Tornadic storms were often categorized with squall lines.
* **Air mass thunderstorm**. Considered to be found across much of the U.S. in the summer.
* **Frontal thunderstorm**. Linear or broken thunderstorms associated with a front.
* **Squall line**. A quasi-linear storm system, generally considered to occur ahead of a front.

7. The severe storm

A severe storm is not necessarily a distinct type of weather phenomena but is simply a thunderstorm that brings enhanced danger to life or property. In the United States, the National Weather Service considers a storm to be severe when it has either a tornado, wind 50 kt (58 mph / 26 m·s^{-1}) or greater, or hail 0.75 in (1.9 cm) or larger in diameter. Canada's thresholds are almost identical but also include a rain rate of 50 mm·h^{-1} (2 inches per hour) or more.

Most severe storms are the result of organized, long-lived convection, which includes many MCS storms and most supercells. Though severe weather is tied with long-lived convection, the *pulse thunderstorm* has been ascribed to severe storms with a very short lifespan. Such an example would be the storm that downed Delta Flight 191 at DFW Airport in August 1985. This storm developed rapidly, produced 70 mph winds, and quickly dissipated. The pulse thunderstorm is not a distinct archetype, however, and should be regarded as any short-lived thunderstorm with an unusually strong updraft or downdraft. The structure may be classified as that of an ordinary unicell thunderstorm but can also include multicell storms. As far as the forecaster is concerned, however, any storm that forms in a very unstable environment or a sheared environ-

Figure 2-12. Weak summertime cell west of New Orleans on 7 July 2008 at 2106 UTC. In the 1960s this would have been called an "air mass thunderstorm", but now we recognize it as a multicell cluster. It is seen here as shown in a 3-D view on GRLevel2 (left) and 10 miles from a Boeing 737 at 35,000 ft (right), both views looking south. Most noticeable is that the radar echo has very little bulk at higher levels, and no strong echoes or echo overhangs are found aloft. The crisp, young towers to the right side of the image became dominant about 30 minutes later as the main cell died. *(Tim Vasquez)*

ment is capable of producing severe weather, and an environment capable of producing pulse thunderstorms can in fact produce long-lived cells!

8. Multicell storms

The simplest model of a thunderstorm involves one cell, consisting of a single updraft-downdraft pair. This is referred to as an *ordinary thunderstorm* or *unicell*, and serves as the building block of the basic thunderstorm model. However modelling and field observation indicates that most updraft areas are composed not of one large, distinctive bubble but of many smaller updrafts — very much an embodiment of the multitude of scales in L. F. Richardson's quote "big whirls have little whirls". Even the simplest type of storm is comprised of an aggregate of updraft bubbles, and this is reflected in the non-symmetrical structure of the cumulonimbus cloud. Needless to say, the long-standing dichotomy between unicell and multicell storms may in fact have little basis, and it could probably be argued that all thunderstorms are multicells.

In practical forecast usage, the term "multicell" is most commonly used for any isolated storm that is neither part of a larger system (an MCS, covered separately) nor contains a persistent rotating updraft (a supercell). There are two basic types of multicells: the cluster and the line.

8.1. MULTICELL CLUSTER. A multicell cluster describes an assortment of storm cells which are in different stages of their life cycle and are not aligned in any particular manner. This is quite common with thunderstorms in weakly sheared environments and with no strong boundaries, such as the weaker storms that frequently occur in tropical regions and in the southern United States.

8.2. MULTICELL LINE. A multicell line is the name given to a storm where different life cycles are in progress simultaneously along the same axis. For example, it is common for the poleward side of the line to be made up of a cell which earlier was mature and is now decaying, while the middle portion is mature and the equatorward portion is in the towering cumulus stage with a "flanking line". In effect, the activity propagates down the line toward the newest cells, and in the storm-relative inflow direction. A multicell line may be seen in severe weather environments where boundaries are not strong enough to create squall lines and instability and shear are not high enough to create supercell storms.

9. Supercell thunderstorms

The term "supercell" was coined by British researcher Keith Browning in 1962. The meaning has gradually grown to encompass the Johns and Doswell 1992 definition of any thunderstorm which contains a deep, persistent mesocyclone. Though the supercell has a fairly persistent updraft and can be very long-lived, the cellular structure of the storm tends to undergo cyclical changes, so the "steady state" character that has traditionally been ascribed to supercells is not entirely correct.

Field research during the 1970s and early 1980s showed that supercells have a range of structures based on the balance between the updraft and downdraft strength. This led to a classification scheme for supercells. A storm with downdraft dominance is categorized as a high-precipitation (HP) supercell, while a storm with updraft dominance is called a low-precipitation (LP) supercell. A storm where neither the updraft nor the downdraft is dominant is known as a classic (CL) supercell. Since updraft and downdraft dominance is not black-and-white but varies considerably from storm to

---

The first known supercell

. . . Four of these cells [on 9 July 1959 in southeast England] became very intense and amalgamated to form a single large cell with horizontal dimensions on the order of 10 mi. [It] maintained a virtually steady state structure throughout a 30-minute period... During the period prior to the development of new cells on its right flank this "supercell" was reaching the greatest heights and intensities of the day as well as producing the largest and most widespread hail.

KEITH BROWNING, 1962
storm researcher

> **Recognizing the RFD**
>
> The proposed supercell model contains an intense updraft and two downdrafts as the main structural features. One downdraft is located in the precipitation cascade region downwind (relative to the 3-5 km AGL flow) of the updraft. The other downdraft lies immediately upwind of the updraft (relative to the 7-10 km AGL flow). It is the upwind or "rear flank" downdraft which is hypothesized to be of critical importance to mesocyclone structure, storm evolution and tornadogenesis.
>
> LESLIE LEMON &
> CHARLES DOSWELL, 1979

storm, the LP-CL-HP types, too, are a spectrum and are not distinct supercell types.

There is a trap in using the visual precipitation intensities to gauge the storm structure. Storms frequently undergo changes in precipitation intensity that may cause an individual storm to appear LPish, HPish, and classic all in just 15 minutes. There is also some criticism of the LP/CL/HP system with one school of thought proposing that supercells should merely be described in terms of inflow/outflow dominance. It should be noted that visual observations of a storm are more forgiving of rain and downdrafts, resulting in a bias toward the LP side of the spectrum, while radar often encounters strong reflectivity of hail and large droplets and demonstrates a bias toward the HP range.

9.1. FORWARD FLANK DOWNDRAFT (FFD). In a supercell, the *forward flank downdraft (FFD)* is equivalent to the main downdraft area of any other isolated storm type, falling downshear of the updraft. The special name used here only serves to identify it as separate from the RFD, to be described shortly. The FFD contains the bulk of the storm's large hail, especially near the updraft. The nature of the FFD's formation is adequately covered in the preceding section titled "Downdraft".

9.2. REAR FLANK DOWNDRAFT (RFD). Early studies (e.g., Nelson 1976) noted that a significant downdraft frequently develops in the rear (upshear) flank of supercell updrafts. This has come to be known as a *rear flank downdraft (RFD)*. Originally it was speculated that this downdraft was caused by barrier flow around the updraft, where the upper-level winds impact the updraft and are forced downward. However, evidence strongly suggests the RFD, like the FFD, develops from evaporative cooling and is accelerated by a downward nonhydrostatic vertical pressure gradient force. Since a supercell updraft usually correlates with a mesocyclone, the RFD is forced to wrap cyclonically around the updraft as it descends. Visually, the RFD is usually manifested by a clear slot (Moller 1974). In some cases it wraps around the updraft, occluding it from the storm inflow.

The RFD has a substantial correlation with a supercell's damaging straight-line winds. Theodore Fujita's earliest studies found

THE THUNDERSTORM 45

Figure 2-13. **LP supercell** in March 1997 in Norman, Oklahoma, looking west. Note that the downdraft and precipitation cascade is on the left side of the image, indicating northwesterly storm-relative flow aloft. *(Tim Vasquez)*

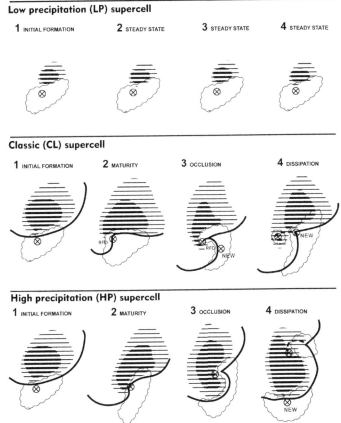

Figure 2-14. **Lifecycles of three key types of supercells**. Note the aggressiveness of the rear-flank downdraft in the high-precipitation supercell type. *(Tim Vasquez)*

that the tornado's rotational damage track was often associated with a track of straight-line microburst winds, sometimes known as "plow winds". A better understanding of supercell structure during the 1970s revealed that this track was associated with the path of the supercell rear-flank downdraft, which in some storms was not only intense but also due to its eastward movement it combined with the forward speed of the storm.

9.3. CLASSIC (CL) SUPERCELL. A supercell in which neither the updraft nor the downdraft are dominant, except perhaps for brief periods, is known as a *classic supercell*. The forward-flank precipitation core is usually opaque with rain and hail. The updraft tends to be fairly large and extensive. Classic supercells are favored by moderate storm-relative anvil-level winds, of about 35 to 60 knots.

The classic storm follows the textbook structure commonly ascribed to supercells, with a hook echo on the left or right flank. However it often goes through cycles in which the downdraft, most

Figure 2-15. Classic supercell near Altus, Oklahoma on 16 June 2008 showing prominent updraft features, including the rain-free base and helical striations. *(Olivier Staiger)*

commonly the RFD, temporarily becomes dominant. This may cause an classic supercell to temporarily take on low-precipitation or high-precipitation characteristics and may lead to misleading conclusions from spotter reports.

The classic supercell produces most instances of long-lived tornadoes. This is because the downdraft is sufficiently active enough to allow for tornadogenesis, but the overpowering downdrafts or weak shear profiles that favor HP modes (to be covered shortly) do not exist.

9.4. LOW PRECIPITATION (LP) SUPERCELL. During the 1970s chase programs it was noticed that some storms were rather devoid of downdraft features and had drier characteristics. These storms rarely showed visual evidence of a strong precipitating downdraft and featured a very symmetrical updraft. Since chase teams usually found these types of storms on the dryline rather than along fronts and outflow boundaries, they became known as "dryline supercells". In 1983, Bluestein and Parks refined this definition into the *low-precipitation supercell*, or LP supercell.

It must be remembered that the LP supercell is not a distinct storm type but rather a storm on the supercell spectrum that has relatively low downdraft dominance. An LP storm usually has a small, semitransparent forward flank downdraft which sometimes appears benign but in fact contains very large hail. The updraft area is usually rain-free. In extreme cases an LP storm may appear to be nothing but a cumulonimbus stovepipe into the troposphere measuring only a few miles in width, and the rain-free base takes on a laminar, bell-shaped, or "mothership" appearance.

Strong storm-relative anvil-level winds, generally over 60 knots, favors LP modes. This causes great dispersion of the forward flank downdraft and substantially eliminates downdraft processes in and near the updraft. Since there is little interaction between the downdraft and updraft in an LP supercell, the updraft tends to have a continuous, persistent structure and the entire storm may be considered a steady state process. Because of this lack of downdraft interaction, the LP supercell almost never produces tornadoes. The steady state nature of the LP storm and the weak dominance of the downdraft also diminishes the risk of severe outflow winds.

Flooding tip

Classic or HP supercells in high-instability environments are always capable of producing torrential rain with the risk of flooding. This is because the moisture flux into these storms is enormous.

48   THE THUNDERSTORM

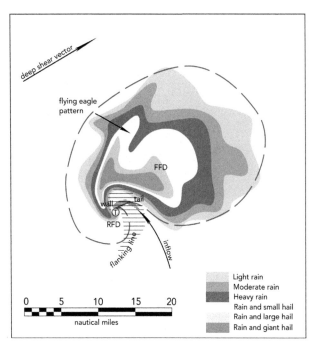

Figure 2-16a. Classic (CL) supercell conceptual schematic. This model diagrams the 3 May 1999 supercell shortly after passing through Bridge Creek, Oklahoma. The boundary of the cold pool is shown with a thick dashed line, and the updraft base is in coarse hatching. *(Tim Vasquez)*

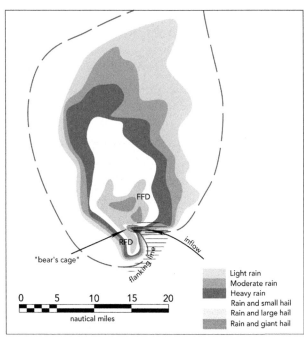

Figure 2-16b. High-precipitation (HP) supercell conceptual schematic. This shows the 5 May 1995 supercell shortly before it produced massive hail devastation in Fort Worth, TX. *(Tim Vasquez)*

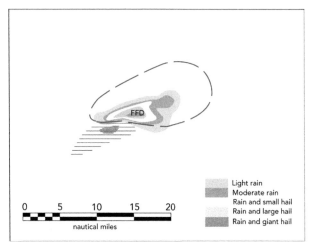

Figure 2-16c. Low-precipitation (LP) supercell conceptual schematic. Based on the 1 May 2008 central Oklahoma supercells. *(Tim Vasquez)*

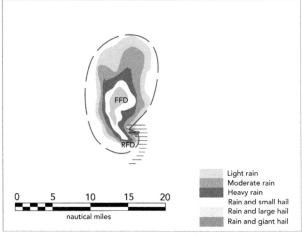

Figure 2-16d. Mini- (low-topped) supercell conceptual schematic. Based on the 10 April 2005 storms in central Kansas. *(Tim Vasquez)*

9.5. HIGH-PRECIPITATION (HP) SUPERCELL. The rainy tornadic storms of the southeast U.S. gained the name "southeast U.S. supercell" but operational forecasting showed that this type of storm to be very common. In fact it is probably the most common type of supercell. In an attempt to present this archetype for spotters the HP supercell was formally defined in Moller and Doswell 1988 as a storm with copious amounts of water in the wrapping downdraft.

Storms which are near the HP end of the spectrum have highly developed downdraft areas and extensive, opaque precipitation shafts. The updraft is found on the inflow side of the storm and is often obscured from view by the rain shafts. In many cases, the only place to view the updraft is in a relatively rain-free area on the inflow side of the updraft, a place called the "inflow notch" or the "bear's cage". The updraft tends to become highly rain-wrapped as it occludes. The storm has a kidney bean shape on radar.

Sometimes a rain-wrapped updraft is mistaken for an HP supercell when in fact it is just a classic supercell going through a rain-wrapped occlusion cycle. Monitoring the storm's life cycle over time is the key for determining where on the LP-HP spectrum a storm lies.

Storms that take on HP modes can be devastating and are capable of producing devastating hailfalls, torrential rain, and flooding. Ironically, tornado production is marginal at this end of the scale because of extensive downdraft activity, which leads to vigorous undercutting and occlusion of updrafts. Mesocyclones are normally weak to moderate and have a relatively large size. HP storms may take on squall line characteristics and be associated with derechos.

Meteorological conditions which are favorable for high-precipitation modes include *weak storm-relative winds at anvil level*, of roughly less than 35 knots. This weak flow inhibits the dispersion of precipitation away from the updraft, causing it to reseed the updraft and improve its precipitation efficiency. *Orientation of boundaries parallel to mean tropospheric flow* tends to cause storms to rain into other storms further down the boundary, known as "seeding", and which forces seeded storms that would otherwise become supercellular to develop HP characteristics. Finally, HP supercells tend to develop in *humid environments* which boost precipitation efficiency of the storm.

Pet peeves

Pet Peeve 1: "Storms that turn more to the right will have greater tornado potential." Storms that move more to the right (in the northern hemisphere) will experience larger SRH in most cases, but that does not equate to increased tornado potential. The most common reason for extreme rightward propagation is the forcing of new updrafts along the rear-flank gust front. Deep, cold outflow (typical of HP supercells) is good at forcing new updrafts, but it is also too stable for significant tornadogenesis. Storm motion is an effect, not a cause.

Pet Peeve 2: "Storms will produce tornadoes when they cross a boundary." Boundaries, such as stalled fronts or outflows, can be a source for increased moisture and low-level shear. However, the thermodynamic properties of the boundary are critical! Moving boundaries still tend to be stable on the "cool" side near the ground, and all the low-level shear in the world will not matter if the layer containing the shear is too stable for thunderstorm updrafts. It is important to look for "older" boundaries where the cool air has modified and moisture has accumulated on the cool side of the boundary.

RICH THOMPSON, 2009
personal communication

Figure 2-17. High-precipitation supercell near Pratt KS on 26 May 2008, showing part of the updraft. These storm types are very dangerous, difficult to chase and spot, and are probably responsible for most severe weather damage in the United States. Chasers hoping to catch a glimpse of the tornado must get near the so-called "bear's cage", partly visible to the right. *(Olivier Staiger)*

9.6. MINI- (LOW-TOPPED) SUPERCELL. Some supercells develop compact structures with low storm tops, usually under 30,000 ft MSL, and contain relatively small mesocyclones. They were originally documented in the context of supercells embedded within hurricanes, but isolated examples on the Great Plains were extensively identified by John Marwitz in 1972 and Jon Davies in 1990. They have become known as either *miniature supercells*, *minisupercells*, or *low-topped supercells*. They can and do produce tornadoes, but limited storm dimensions above the freezing level tends to impede large hail production. It is important to remember that because of the compact structures, beam width problems become more acute, making typical severe weather signatures more difficult to detect. Consequently forecasters should bias the forecast towards a higher likelihood of tornadoes or high wind when analyzing a mini-supercell on radar.

Mini-supercells do show a strong correlation with high-shear environments. Large amounts of instability are not a prerequisite. Markowski and Straka 2000, for example, demonstrated that mini-supercells can develop with as little as 300 J·kg^{-1} of CAPE. This makes the type a major forecast concern in regions of the world with cool climates, especially where the storm may not be easily recognized on radar.

9.7. ANTICYCLONIC SUPERCELLS. In a splitting environment with relatively straight hodographs, an anticyclonic supercell, a left mover may continue to persist without dissipating. It usually has a mirror-image structure of a textbook classic or LP supercell. They are known to produce long swaths of giant hail. Overall, the anticyclonic split is not a distinct storm type except for its sign of rotation. Its structure is adequately covered in the multicell and supercell sections, except considered as a mirror image of the typical right-moving (NH) member, and its genesis is discussed in the section on splitting.

10. Heavy rain and flooding

Heavy rainfall rates comprise another type of severe weather and are enhanced by strong updrafts, high water content of the updraft air, and collision and coalescence being the dominant method of droplet growth.

10.1. SLOW STORM MOTION. Slow storm motion is a predictor for heavy precipitation. Not only steering flow should be examined but the propagation vector must be considered, too. With respect to the ground, weak wind profiles are correctly assumed to be associated with slow motion, but *slow storm motion is also possible with strong winds aloft*. The key is to examine the hodograph. It is extremely easy to construct plausible hodograph plots which show strong speeds and large directional changes with respect to the ground but which show stationary storm motion, especially when the propagation vector is factored in. In weaker flow storms may also propagate toward favored orographic features, remaining stationary over those locations.

10.2. CELL TRAINING. Training is the repeated passage of new convective cells over a given location. Line movement which is parallel with an initiating boundary may cause this to occur, as might be caused by the synoptic pattern in an "anafront" situation. A unidirectional shear profile can be conducive to training since this implies that mean tropospheric flow is roughly parallel with the orientation of surface boundaries.

10.3. STRONG UPDRAFTS. Without a strong updraft, the storm is inefficient at transporting low-level water vapor into the upper part of the storm and converting it into precipitation. The factors that enhance strong updrafts are outlined in the earlier sections. A strong updraft allows for longer residence times of precipitation particles in the cloud.

10.4. HIGH SPECIFIC HUMIDITY. High specific humidity means that the mass of water vapor in a given volume of air is high. Inflow air with high specific humidity can be determined by assessing the inflow mixing ratio or the inflow dewpoint. Since dry air entrainment into an updraft diminishes the specific humidity within a storm cloud, high specific humidity is enhanced by wide updrafts and moderate to high relative mid-level relative humidity. High humidity aloft also diminishes the effect of evaporation in precipitation cascades. The presence of many thunderstorm cells in the same cluster can also add significant amounts of moisture to the troposphere, curbing the effect of dry air.

10.5. EFFICIENT PRECIPITATION PRODUCTION. Collision and coalescence of liquid droplets is much more efficient for heavy precipitation production than for ice processes. Therefore a very high freezing level, as is seen in warm season episodes, usually corresponds to a deeper "warm" cloud where collision and coalescence is dominant. Furthermore, processes must exist which allow water droplets to quickly collide: this requires a broad spectrum of droplet sizes, all of them with different fall rates.

10.6. CELL SEEDING. Seeding of the updraft by neighboring sources, such as upstream flanking lines or upstream storms, can improve the precipitation efficiency of the storm. This introduces different particles with different sizes into the cloud, enhancing collision and coalescence and other growth processes.

11. Cloud structures

In this section we consider the visual element of storm structure. Though desk forecasters may not have much interest in interpreting visual structure, spotter, media, and mobile reports are providing increasing amounts of real-time storm information to supplement radar data. Being able to make assessments of cloud observations reveals a vast amount of information about the storm.

11.1. WALL CLOUD. Wall clouds are pronounced, persistent lowerings which were first documented in an analysis (Fujita 1960) of the 1957 Fargo tornado. Wall clouds with a flared shape have also been described as pedestal clouds (Marwitz et al. 1972).

The wall cloud forms primarily under the storm's rain-free base near the updraft-downdraft interface. It may have an accompanying tail cloud pointing in an eastward direction along the leading edge of the outflow. The wall cloud may show rotation and in many cases if this rotation becomes violent it will correspond to where the tornado develops.

The wall cloud is believed to be caused by air parcels close to the downdraft (or forward flank downdraft, in the case of a supercell) which are chilled and humidified by precipitation falling into the parcels and partially evaporating. This air, because it is cooler and has not lost any moisture, has a lower dewpoint depression. Thus when it is lifted back into the updraft, it needs a relatively small amount of lift, compared to pure inflow, before it condenses into a cloud.

It is speculated that one factor that affects the amount of wall cloud lowering may be the humidity of the inflow air. Evaporative cooling is stronger in dry air, and this can significantly change parcels close to the downdraft, producing low wall clouds. On the other hand, evaporation is not very efficient in humid air, thus parcel characteristics do not change much even with exposure to precipitation, and marginal wall clouds are produced. The ambient air temperature, the temperature of the precipitation droplets, and

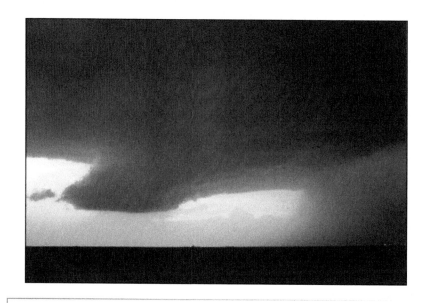

Figure 2-18. Wall cloud and forward flank downdraft area on 2 June 1995 near Lubbock, Texas, looking westward. The large cloud mass above is the storm's rain-free base. *(Tim Vasquez)*

especially the exact inflow trajectory can also be significant factors in wall cloud character.

Not all tornadic storms produce wall clouds, and not all wall clouds are associated with tornadoes. Furthermore the link between pressure deficits in the mesocyclone and the formation of wall clouds is vague and controversial. Overall, the existence of a wall cloud is not a reliable indicator of tornado development and any inferences should hinge on its rotation.

Spotter training in the 1980s heavily emphasized wall clouds, and since they don't always occur and other processes within the storm are equally important, there has been a consistent effort since then to reduce this focus. Wall clouds are only one small feature of the thunderstorm. Consistent observations of storm structure, rotation, and outflow can yield much more information about the potential for tornadogenesis.

11.2. TORNADO. *The tornado is described separately in Chapter 4, Tornadoes.*

11.3. SHELF CLOUD. The shelf cloud term was initially used to describe the updraft base (Fankhauser 1976). It has come to describe a smooth, wall-like appearance of the leading edge of the updraft, and often is used to describe arcus clouds (see below) which are connected to the storm cloud. The shelf cloud may take on a terraced appearance (Figure 2-19) if stratified layers with varying moisture properties are ingested into the updraft. These all have different lifted condensation levels, resulting in multiple cloud base heights.

11.4. ARCUS OR ROLL CLOUD. A dense, horizontal roll with more or less tattered edges, situated on the lower front part of certain clouds and having, when extensive, the appearance of a dark, menacing arch (WMO 1956). Roll clouds have been linked with storms as far back as the 1920s.

Roll clouds can outlast the storm by many hours and travel long distances. They are responsible for the spectacular morning glory phenomenon that occurs along the north coast of Australia, particularly in the Gulf of Carpentaria.

11.5. SCUD CLOUD. Disorganized shreds of stratus underneath or near the storm base are referred to as scud clouds. They are associated with high relative humidity in rain-cooled precipitation areas. In rainy areas they can solidify into a broken or overcast layer only a few hundred feet above the ground. Scud clouds have no real significance but are sometimes mistaken for wall clouds and tornadoes.

11.6. MAMMATUS. Mammatus is the common name for pouchlike bumps on the underside of a storm anvil plume. Thus their height is typically about 20 to 30 thousand feet above sea level. They are not just confined to thunderstorms, however; they are sometimes observed on the underside of dense altostratus and even stratocumulus decks. Coloration is relat-

Figure 2-19. A thunderstorm with a multi-tiered shelf suggests lift of poorly-mixed layers with different condensation levels. These multi-tiered decks can be spectacular on squall lines. 10 August 1986, Garland, Texas. *(Tim Vasquez)*

56   THE THUNDERSTORM

ed to illumination, with orange and red colors resulting from dusk illumination and gray and bluish colors originating from cloud and sky illumination.

Their association with descending ice crystals has been well understood since the late 19th century, but mechanisms at the cellular level has not been well understood. It should be noted that during the 1950s and 1960s some authoritative sources linked tornadoes to mammatus, particularly in pilot training material. This persisted as a troublesome myth well into the storm spotter era. Mammatus and tornadoes develop in entirely different parts of the storm at completely different altitudes, and are not related. ¤

Figure 2-20. Mammatus clouds under an extensive anvil region. Photographed near Vernon, Texas on 1 June 1990, looking southwest towards the supercell which at the time was producing the Bakersfield Valley tornado. *(Tim Vasquez)*

3

MESOSCALE CONVECTIVE SYSTEMS

A most terrific thunderstorm came up last night; the thunder tumbled from the sky, crash upon crash, as though all was being rolled together like a scroll; the fiery chains of lightning streaked the heavens from zenith to horizon. The rain came in torrents, and the wind blew almost tornadoes. Our cabin, seemingly, was but little security against its wildness. When we heard the storm approaching, we dressed ourselves, wrapping Indian blankets about us, and made ready to protect our children from the rain that was then dripping through the roof. We put all our bedding around them, and all we could see to get by the glare of the lightning (could not keep a candle lit), spread our umbrellas, (five in number,) placed about, and held over them. We all got wet, and were obliged to lie in our wet beds till morning. This morning all was calm; the bright sun ascended up into a cloudless sky, as majestically as though there had been no war in the elements through the night. But the rain had dissolved our mud chinking, and the wind had strewed it all over and in our beds, on our clothes, over our dishes, and into every corner of the house. Have had all our sheets to wash, beds and blankets to dry in the sun and rub up, our log walls to sweep down, our shelves and dishes to clean, and our own selves to brush up. "Such is prairie life," so they say.

— Miriam D. Colt
Went to Kansas, an account of the failed commune
of Octagon City in southeast Kansas
June 3, 1856

Long before supercells were understood, squalls accompanied by thunderstorms were a common occurrence in the Mediterranean and Atlantic and were greatly feared by mariners. The word "squall" dates to at least 1690, referring to a sudden, violent gust of wind, often bringing chaotic weather. With the dawn of the 20th century, meteorologists began using the term "squall line" to describe linear bands of thunderstorms over continental regions. With the advent of mesoscale meteorology in the 1980s, these large areas, organized of thunderstorms were given context and became known as *mesoscale convective systems*.

1. MCS types

Though it is convenient to think of anything larger than a supercell as a simple squall line with two-dimensional flow, this is not necessarily true. Severe weather is primarily the result of significant three-dimensional circulations. The term "mesoscale convective system" helps cover all such systems. An MCS may not necessarily be a simple squall line: it can be a large cluster of summertime storms, a giant bow echo, or even a hurricane!

The key attribute of an MCS is a common cold pool. This cold pool is caused by cold outflow air at the surface. While cell development may occur over the cold pool, most of the strong ascent occurs along the edges of the cold pool. For this reason, MCS storms often evolve into a somewhat linear configuration.

1.1. SQUALL LINE. A squall line is a generic word for an MCS with cells along a common axis. Ideally the squall line has a two-dimensional structure where any given slice orthogonal to the line reveals the same basic cross-section. However this is rarely the case, as individual constituent cells are often evident in squall lines and the "discreteness" of cells may lie anywhere along a spectrum from being deeply merged with one another with individual cells difficult to discriminate to broken squall lines where individual cells are highly evident. Discreteness reduces cell competition and allows for transverse circulations, increasing the risk for severe weather.

Interestingly in the late 1940s, the Thunderstorm Project studied 27 squall lines and came to the conclusion that squall lines either developed in benign locations in the warm sector ("air mass

---

The squall line, circa 1950

There is considerable disagreement regarding the cause of squall lines. One of the older theories is that the squall line is the result of a cold front aloft. A difficulty with this explanation is the frequent lack of any evidence of a front aloft prior to the formation of the squall line.

Harrison and Orendorff examined a number of possible factors and found that the only ones applicable in most cases were (1) a "pseudo-cold front" and (2) convergence into a trough ahead of the real cold front. In a later United Air Lines report it is concluded that the most acceptable explanation is that a mass of rain-cooled air sets up a shallow local cold front in the warm sector, which subsequently releases the thunderstorm energy ahead of the real cold front. This is essentially in agreement with the opinion stated in the final report of the Thunderstorm Project.

C. W. NEWTON, 1950
University of Chicago researcher

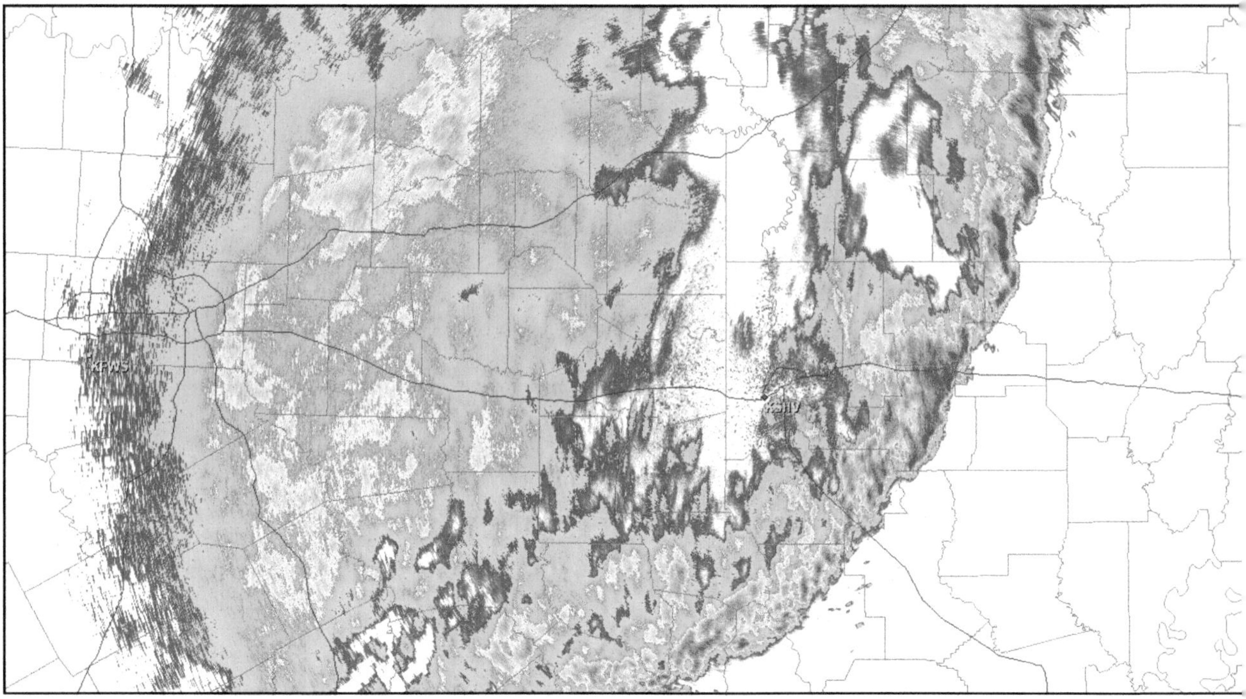

Figure 3-1. Mesoscale convective system in Louisiana on 9 October 2009 showing an extensive trailing stratiform area. On the lower right of the page the MCS has squall line characteristics but closer inspection shows numerous discrete, localized cells. In the upper right, stronger upper-level support has allowed the development of individual bow echo storms.

squall lines") or along or just ahead of a cold front ("pre-frontal squall lines"). At this time the squall line was also believed to be responsible for most tornadoes. By the 1950s new research (Brunk 1953) cast doubt on that long-standing idea, and now it is believed only about 18% of tornadoes are associated with squall lines (Trapp et al 2005).

1.2. QUASI-LINEAR CONVECTIVE SYSTEM (QLCS). The quasi-linear convective system (QLCS) (Weisman and Trapp 2003) refers specifically to a squall line which is *highly two-dimensional*, where cell discreteness is very low to nonexistent. This term was introduced to help provide a framework specifically for squall line tornadoes which do not involve classic tornadogenesis and deep, persistent mesocyclones. A QLCS is not a specific type of storm, however, and only covers squall lines which are exhibiting a very high degree of linear structure.

1.3. MCS CLUSTER. A large MCS without a dominant linear structure is generally referred to simply an MCS, rather than as a squall line. There is not really a term that describes it, though "storm cluster" may be suitable in some instances. In any case, this type of MCS sometimes occurs in weak summertime situations and in regions of widespread isentropic lift north of a warm front, and is usually fairly weak. The structure shows an amorphous mass of precipitation with embedded convective elements. Flooding may be a significant hazard in some cases since clusters are favored in weakly sheared environments with slow cell motion.

1.4. MULTICELL CONVECTIVE COMPLEX (MCC). The multicell convective complex, or MCC, is a special type of MCS defined *only by its structure as observed by satellite*. According to Maddox 1980 for an MCS to be an MCC the -32 deg C cloud top covers over 100,000 km^2, the -52 deg C cloud top covers 50,000 km^2, and these requirements persist for at least 6 hours. Given the forecast tools currently in use, there is not really a practical way to make such determinations, so it is possible the definition of the MCC may evolve in the decades ahead. The classic MCC is associated with much of the warm-season rain that occurs in the Midwest region, and because of its location, season, and precipitation yield, MCC events have great influence on United States corn and soybean production.

1.5. DERECHO. A derecho is actually not a specific type of MCS but is a regional-scale windstorm event produced by a strong MCS, which may be comprised of a squall line, bow echoes, or even supercells sharing the same cold pool. The derecho will be discussed in detail later.

2. MCS features

As far back as the 19th century, cold, gusty outflow air was recognized as a defining characteristic of the squall line. This very much holds true today, with the cold pool and gust front shaping the behavior, propagation, and severity of the system. The MCS also has a number of other important features which are described below.

Squall lines on radar

In 1985 Bluestein and Michael Jain reviewed 40 Oklahoma squall line events in the 1970s associated with severe weather and divided them into four radar-observable categories. These classifications are occasionally encountered in descriptions of MCS types.

- **Broken line**. The formation of a discrete line of cells which transforms into a solid line. This was found to be associated with most of the Oklahoma cases investigated by Bluestein and Jain.

- **Back building**. Signified by the periodic appearance of new cells upstream. This was also a common mode of development.

- **Broken areal**. The development of an amorphous area of moderate-to-intense cells into a solid line of convection.

- **Embedded areal**. A convective line appears within an area of stratiform precipitation.

Key forecast concepts

* The trailing stratiform (TS) mode is favored when there is strong low-level inflow and relatively weak anvil-level outflow.

* The leading stratiform (LS) mode is more common in lines with weak low-level inflow and relatively strong anvil-level outflow.

2.1. COLD POOL. A large pool of cold air is the primary characteristic of an MCS. The edge of the cold pool is known as an outflow boundary or gust front, and it continuously forces the lift of new cumulonimbus towers along a linear axis. The cold pool is largely maintained by the stratiform precipitation area (to be discussed shortly).

2.2. UPDRAFT TOWERS. Cumulonimbus towers form along the edge of the cold pool, usually where low-level convergence is strongest. The updraft towers are composed of high theta-e air. These aggregate into a very large, extensive line that parallels the outflow boundary. The area underneath the updraft towers tends to be relatively rain-free and may contain a menacing sky, arcus clouds, and lightning. Brief wall clouds are common.

2.3. DOWNDRAFT. Immediately behind the gust front, a strong downdraft is present, which often appears as a wall of precipitation. The highest winds in the MCS are usually found along the leading edge of this downdraft core, and the highest surface pressures are found directly underneath it, where it produces the highest barometric pressures found within the MCS's mesohigh. The downdraft at times may be augmented by the rear inflow (to be discussed shortly).

2.4. STRATIFORM PRECIPITATION AREA. An MCS is often, but not always, associated with a large area of continuous rain measuring up to 100 or 200 km in width. It may be separated from the active downdraft area by a transition zone with little or no precipitation. Occasional lightning activity, particularly in-cloud lightning, may be observed in the stratiform area with the characteristic long, rolling thunder. This stratiform area may also show a bright band on radar: a horizontal layer aloft where reflectivity is very high due to the melting of ice and snow. The stratiform precipitation area is comprised primarily of a low-level downdraft and it helps reinforce the cold pool. Low stratus or "scud" layers at an altitude of less than 1 km AGL are common within the stratiform area and frequently obscure the view of the anvil canopy overhead.

When the stratiform precipitation area trails the MCS, it is known as *trailing stratiform* (TS) configuration, sometimes called

MESOSCALE CONVECTIVE SYSTEMS 63

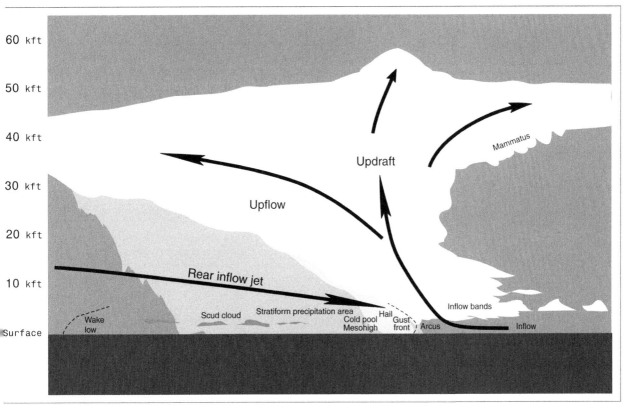

Figure 3-2. Conceptual cross section of a trailing-stratiform MCS. Typically this would be looking northward in either hemisphere. Significant precipitation is shaded gray, with heaver precipitation and hail in light gray. The updraft area is driven largely by gravitational convection, while the upflow region consists of decaying updrafts and slantwise convection. *(Tim Vasquez)*

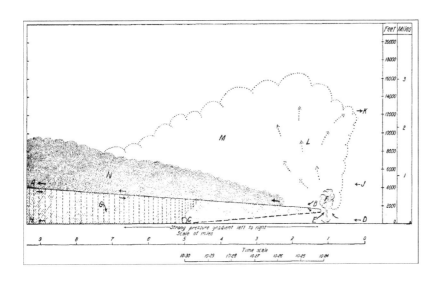

Figure 3-3. Published in 1928, this conceptual model of a squall line was remarkably accurate for its time. It synthesized the Norwegian cyclone model with cloud observations and "theoretical considerations". The feature at F is not a tornado but is identified as the "characteristic long roll cloud of the squall line". *(M. A. Giblett, 1928, "M. A. Giblett on Line Squalls", Mon. Wea. Rev., **56**, 7-11)*

64 MESOSCALE CONVECTIVE SYSTEMS

Figure 3-4. Cross section of damaging squall line northwest of Nashville TN on 16 June 2009 at 1856 UTC.

Right: A standard base reflectivity product is shown, along with the cross-section segment.

Center: A vertical cross-section of storm-relative velocity, with isotachs added for legibility purposes. Positive values (solid lines) indicate flow away from the radar, westward, and negative values (dashed lines) indicate flow toward the radar, eastward. Since the line is moving toward the radar and the segment is parallel with the beam, most horizontal motions should be in this plane, though the radar of course cannot detect vertical motion and so the 3-D structure here is presumed. However the updraft/upflow area and rear inflow regions are quite visible.

Bottom: Same velocity markings as viewed against a reflectivity cross section.

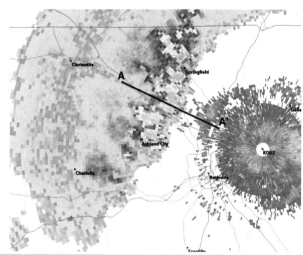

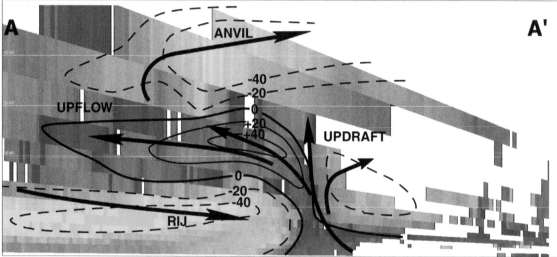

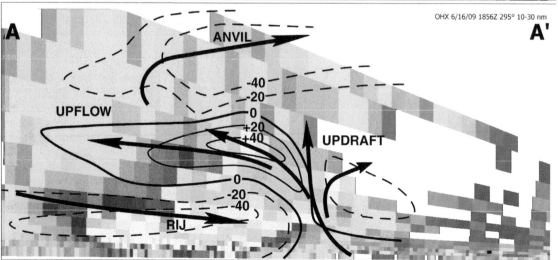

leading-line-trailing-stratiform (LLTS). This is favored when the system-relative upper-level wind component is from the front to the rear of the system. It is the most common configuration and is forward-propagating, moving faster than the mean flow. When stratiform precipitation falls downwind of the MCS, it results in a *leading stratiform* (LS) or trailing-line-leading-stratiform (TLLS) system. It is favored when the system-relative upper level wind component is from the rear to the front of the system. The system propagates rearward and so the system tends to move more slowly than the mean upper flow. As a result the primary threat is flooding and "training" of cells over specific locations. Finally storms may take on a *parallel stratiform* (PS) configuration where precipitation falls on the MCS itself, particularly further up the line where it forms a very broad core. The line is usually oriented perpendicular to the upper-level winds, as what might occur with a stationary front.

Descending rear inflow jet
A descending RIJ tends to be associated with weak shear in the lower troposphere and/or weak CAPE.

2.5. UPFLOW. As the updraft towers mature, they are often undercut by the cold pool, with many older, updraft cells above the pool. This region comprises a large-scale area of weak ascent with embedded, dying updraft towers. Aside from these localized areas of weak convective ascent, the dominant convective mechanism in the upflow area is primarily moist symmetric instability (q.v.) above the underlying cold pool. Overall, this upflow area is reinforced by the large-scale release of latent heat due to condensation and freezing processes. Over time this results in a large horizontal circulation that may develop into a mesoscale convective vortex (to be discussed shortly). *Note: In this text we refer to this area as "upflow" (1) for simplicity; (2) to avoid specific terms like "ascending FTR" or "ascending RTF"; (4) to avoid words like "updraft" which should identify individual buoyant updrafts; and (4) to avoid further ambiguous use of the term "inflow".*

2.6. MID-LEVEL LOW PRESSURE. The low-level cold pool and the upflow area tend to move apart from one another, removing mass between them. This produces a low pressure area between the two. Though this mid-level low is not detectable at the surface and produces no observable weather directly, it influences other circulations such as the rear inflow (to be discussed next).

MCS experiments
Listed here are some noteworthy field experiments which have involved squall lines:
* **PRE-STORM**. A 1985 experiment using ground and space based instruments.
* **BAMEX**. A 2003 experiment which used WSR-88Ds in the central United States to study MCS behavior.

Figure 3-5. Mesoscale convective vortex captured 08 July 1997 at 1545 UTC over western Missouri on visible satellite imagery. This is the dying remnant of an MCS that moved southeastward through Nebraska and eastern Kansas. It formed in the pre-dawn hours behind the squall line, which has moved to northwest Arkansas. In spite of the ominous appearance of the MCV, its circulation was weak and only light showers were occurring in the spiral bands. *(Tim Vasquez / McIDAS software / NOAA data)*

2.7. REAR INFLOW. In the more common trailing stratiform configuration, mid-level air (at about 2 to 6 km AGL) moves through the rear area of the MCS in a rear-to-front (RTF) direction. This is called *rear inflow*. It is not inflow in the sense of feeding the updrafts with buoyant energy but rather it is a conveyor that flows into the back of the MCS.

How does the rear inflow jet form? In the updraft area, latent heat is being released, accelerating parcels upward, while down below, air in the downdraft area is accelerating. Due to each of these processes, mass is being removed from the mid-levels of the atmosphere. This produces a mid-level low pressure area. The rear-inflow jet is a response to these pressure falls. Quite often the air at this level is already moving west-to-east through the storm, and given the acceleration due to the mid-level low it intensifies. The greater the positive and negative buoyancies within the storm, the more intense the mid-level low and in turn the stronger the rear-inflow jet.

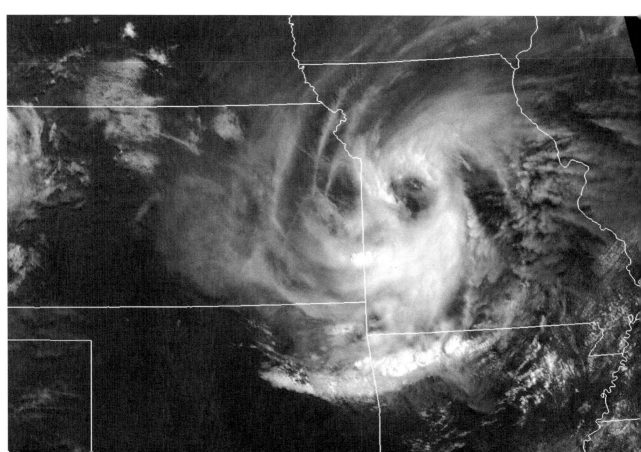

In many cases, a rear-inflow jet descends to the surface just behind the line of cumulonimbus towers. This ducting is caused by convergence of the rear-inflow jet with the buoyant, rising air on the leading edge of the line. This may bring high-velocity air to the surface, i.e. cause the downward transport of momentum. However other RIJs remain elevated aloft without descending, becoming what is known as a non-descending RIJ. This latter type is more likely to sustain a long-lived MCS.

It is also believed that the rear-inflow jet may capture ice from the aft sections of the upflow area, which has spent time growing into snow flakes and begins falling. Since the rear-inflow jet is continuously infused with dry air, this promotes evaporational cooling of the rear-inflow jet and may help reinforce the strength of the cold pool.

In trailing-stratiform MCS storms, which have a mirror structure of the leading-stratiform system, the role of the rear inflow is assumed by a *leading inflow jet*, which blows front-to-rear (FTR). Since it the leading stratiform configuration is rarely associated with damaging weather this book will focus primarily on trailing stratiform systems which contain rear inflow jets.

2.8. WAKE LOW. In the wake of the MCS and the cold pool, a wake low is found at the surface, which may appear as a trough on mesoscale analyses. It is a reflection of the release of latent heat within the large stratiform region overhead and adiabatic heating of the descending rear inflow.

2.9. PRE-SQUALL LOW. Adjacent to the MCS in the high theta-e (warm moist) air, a pre-squall low may be found. It tends to be weak and may owe its existence to adiabatic warming in the anvil canopy overhead.

2.10. MESOSCALE CONVECTIVE VORTICES (MCV). Beta-mesoscale cyclones of 50 to 300 km in diameter and lifespans of many hours have been identified in some MCS storms. They were first documented by Edward Johnston in 1982, named mesoscale vorticity centers (MVCs), and have come to be known as mesoscale convective vortices (MCVs). They may outlast the MCS itself, appearing as a sort of residual circulation. The MCV quite often is readily ap-

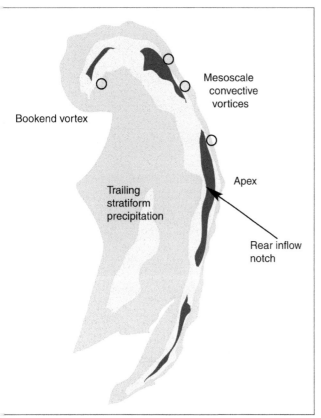

Figure 3-6. Conceptual model of severe weather features in an trailing stratiform MCS. Shading indicates either rainfall rate or radar reflectivity, with enclosed shapes containing higher intensities than the surrounding region. *(Tim Vasquez)*

parent on visible and infrared satellite imagery, showing a cyclonic pattern in the clouds, and surface data may reveal the existence of a meso-beta low.

MCVs are thought to originate from widespread latent heat release within the storm system. This decreases density of the air mass and consequently results in pressure falls and a tendency for a weak cyclonic component in the wind field. Essentially the MCS develops warm-core system characteristics, much like that in a tropical cyclone. The circulation is strongest in the mid-levels, within the stratiform rain area and above the cold pool, and may extend to the surface as a wake low. The MCV may also appear to couple with a cyclonic, decaying bookend vortex (to be discussed shortly). Studies of MCV environments have shown that they are more likely in environments with weak bulk shear and moderate to high instability.

As far as what effect an MCV has on the forecast, these features do serve as a focal area for convective redevelopment, so their movement should be tracked carefully on mesoscale analyses using all available tools. Models tend to handle them erratically due to radiosonde contamination in MCS environments and marginal spatial resolution of the radiosonde network. Therefore it's essential for forecasters to monitor and track MCVs once they're detected. As a basic rule of thumb, considering the shear profile of the tropopause, convection is more likely downshear of an MCV and less likely upshear.

3. MCS dynamics

One important consideration in MCS development is the balance between two horizontal vorticity mechanisms: environmental vorticity and baroclinic vorticity. Environmental vorticity is generated by the shear between low-level and upper-level flow. In a typical atmosphere where strong westerly winds occur aloft, this produces positive horizontal vorticity. This type of shear causes

MESOSCALE CONVECTIVE SYSTEMS

Figure 3-7. Vaulted region behind an outflow boundary which has gusted ahead of the storm, looking east. This is a sign of an outflow-dominant MCS. Photographed 14 April 1999 near Chandler, OK. The photographer had parked alongside the road near a rural driveway and the car was being circled by four cranky dachshunds and pit bulls from a nearby trailer. *(Tim Vasquez)*

the developing cumulonimbus towers to lean downshear (i.e. to the east) with rain falling downshear.

However the MCS is structurally different because it occurs along the leading edge of a cold pool. There is a circulation here because the cold air has a tendency to sink and undercut the warm air, while the warm air has a tendency to rise and override the cold air. This forms a circulation which manifests itself as baroclinic vorticity. Along the leading edge of the cold pool, this has a negative sign. The cold pool can be strengthened by downdrafts that entrain dry mid-level air, enhancing evaporational cooling.

With positive environmental vorticity and negative baroclinic vorticity coupling with each other, the result is that updraft towers which form on the boundary are balanced and tend to rise vertically. This forms the basis for *RKW theory* as developed by Rotunno, Klemp, and Weisman. In a developing cluster of thunderstorms with sheared towers and cold pools beginning to develop, cells that exploit the balance between baroclinic and environmental vorticity are favored for development and this leads to an balanced MCS.

> **The squall line, circa 1894**
>
> The violent outrushing of cool wind that commonly precedes our stronger thunderstorms has been regarded by some observers as contradicting the belief in the convectional [upward] character of the storm as a whole; but when the subordinate dimensions and position of the squall with reference to the other parts of the storm are properly perceived, this contradiction entirely disappears. The [wind] squall does not reach to a great height above the surface of the earth; its upper limit must be below the dark lower front edge of the storm cloud — perhaps half a mile or less above the ground — for there the wisps of the cloud front demonstrate an inflow with respect to the storm in a most unmistakable manner.
>
> WILLIAM M. DAVIS, 1894
> Harvard University scientist

Unbalanced vorticity does not favor strong MCS events. In situations with weak shear and strong cold pool production, the environmental vorticity sign is weak or even negative. The baroclinic vorticity is dominant, and the result is a weak MCS with cells that are rapidly undercut or which mature above the cold pool. These storms are cold-pool dominant. Likewise, in strongly sheared environments but weak cold pool production, the baroclinic vorticity sign is weak. This favors precipitation cascades that fall downshear, and produces discrete storms or weaker leading-stratiform lines. These storms are shear-dominant.

4. Bow echo storms

During the 1970s, it was recognized that when a segment of a squall line accelerated ahead of the line axis, it formed a radar configuration known as a line echo wave pattern (LEWP). Fujita in 1978 also recognized the potential for bowing echoes to produce significant damage. By the 1990s the *bow echo* became recognized as the primary cause of severe weather in MCS situations and a major contributor to derecho events.

The bow echo is a linear storm measuring about 20 to 100 km in length whose center section accelerates forward. It initially develops within an MCS, but may occasionally evolve from a supercell, especially one in which the rear-flank downdraft is much more active. The bow echo takes on its characteristic bow shape which points roughly toward the direction of movement. One side of the bow echo is cyclonic and the other is anticyclonic, similar to the rotation experienced by the hinges when a restaurant waiter pushes open a set of double doors. The cyclonic portion becomes dominant while the anticyclonic part diminishes, so the bow echo eventually takes on a comma shape.

Early modelling (Weisman and Klemp 1982) clearly demonstrated that bow echoes are favored in *high bulk shear, high instability* environments. A strong boundary and a pre-existing MCS are also favored precursors for bow echo development.

4.1. REAR INFLOW JET (RIJ). The bow echo is caused by very strong rear-inflow. With this kind of strength it is called a rear inflow jet

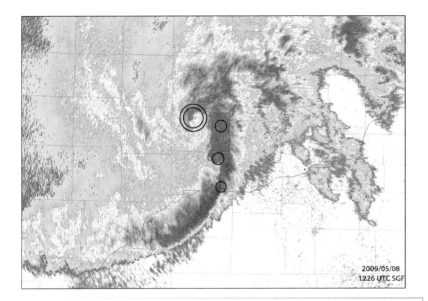

Figure 3-8. Derecho in Missouri on 8 May 2009 showing mesoscale vortices (circles) as detected by WSR-88D velocity products. The concentric circle shows a bookend vortex. (Tim Vasquez / GRLevel2 software <grlevelx.com>)

(RIJ). As with a weaker rear inflow region the RIJ can be enhanced by precipitation loading and evaporational cooling.

The presence of a RIJ can be detected on radar through three key signatures. First is the mid-altitude radial convergence (MARC), a zone of mid-level convergence that indicates the development of the RIJ. As the jet arrives at the surface, a rear-inflow notch (RIN) may form, which is an area of weak reflectivity behind the bow echo apex that is caused in part by the localized descent of relatively clear air. The last stage in the development of a rear inflow jet is the actual bowing of the echo itself. These concepts are discussed in detail in the radar chapter.

4.2. LINE-END VORTICES. Storms which "bow out" in a forward direction show a curl at each end, with the poleward end of an eastward-moving system being cyclonic and the equatorward end being anticyclonic. These are known as *line-end vortices* or *bookend vortices* (Weisman 1993).

The line-end vortices are strengthened by tilting of environmental vorticity downward by the rear-inflow jet. In a typical positively sheared environment this produces cyclonic vorticity to the left of the RIJ and anticyclonic vorticity to the right. Likewise, the tilting of negative baroclinic vorticity upward by the updraft may augment line end vortex development.

MCS storms and shear

By 1985, a landmark study by Joseph Klemp, Richard Rotunno, and Morris Weisman confirmed the idea that MCS systems were highly dependent on environmental shear profiles. The research showed that the strongest lines had strong shear vectors relative to the line in the low levels, but weak ones at higher levels.

Some historical derechos

* July 4, 1969. A derecho. It moved southeast from Lake Erie into Ohio and Pennsylvania, killing 3 who were watching fireworks events from boats along the shore.

* May 4-5, 1989. A progressive derecho moved at 80 mph from Amarillo to Houston, causing millions of dollars in damage in north and central Texas.

* July 18, 1991. A derecho moved from Minnesota into the woods of far west Ontario, causing a large forest blowdown. The strong winds videotaped by an amateur near Pakwash gave rise to the colloquialism "Pakwash storm".

* March 13, 1993. Associated with the Storm of the Century, this derecho swept through the Gulf Coast states and Cuba.

* July 4-5, 1999. A progressive derecho moved from northern Minnesota to Maine in 12 hours, causing massive forest blowdown and over $100 million in damage.

* May 8, 2009. A derecho moved from southwest Missouri into the Mississippi River valley during the day, with several deaths reported.

Due to the contribution of the Coriolis force, the cyclonic vortex tends to become dominant and the anticyclonic vortex dissipates. This eventually gives the bow echo a comma shape. Since the line-end vortex is centered in the downdraft region, it is rarely associated with tornadoes. However it is capable of coupling with the rear-inflow jet and strengthening it.

4.3. LEADING EDGE MESOSCALE VORTICES. Modelling work by Trapp and Weisman demonstrated the ability for mesoscale convective systems to produce gamma-mesoscale (2-20 km) vertical vortexes along the leading edge of the outflow boundary. The favored area for these vortices are poleward of the bow echo apex. These vortices develop from horizontal vorticity along the outflow boundary augmented by thermal contrast. These are then tilted into the updraft. While at fine scales vortices of this type produces gustnadoes, this scale of vortex is associated with mesocyclones and tornadoes. More on this is discussed in the tornado chapter.

5. Derecho

The derecho (pronounced de-RAY-show) is not a type of thunderstorm but a convective *wind event* resulting from a family of downburst clusters, usually from one or more bow echoes or even a series of damaging supercells. The term "derecho" was used for convective windstorms during much of the 1890s, then died out. until it was resurrected in 1983 by SPC forecaster Bob Johns. The National Weather Service defines an MCS as a band of storms that has 50 kt or greater winds along its entire length. The path width is usually greater than 40 miles. Derecho storms occur primarily in the eastern half of the United States, with exception of the Gulf Coast region. A 1998 study by Bentley et al identified the Midwest and lower Plains states as being the key derecho regions, with two general hot spots: one in Oklahoma and the other in the region south of Lake Erie.

5.1. CONFIGURATION. The derecho is indicated on radar by the appearance of a bow echo, measuring in size from roughly 20 km to 200 km and lasting for many hours. A derecho occurs on a spectrum between two types: *serial*, consisting of a line consisting of small bow echoes or HP supercells; or *progressive*, with the entire line forming a bow shape. The serial mode is most common when the mean tropospheric wind is relatively parallel to the squall line, rather than perpendicular to it. Tornado production is more likely in the serial derecho since the broken line structure is more favorable for the generation of mesovortices, whereas inflow into a progressive derecho is more symmetric and not as prone to break up into vortices.

5.2. Derecho Mechanisms. The main mechanism for damaging MCS winds is when strong outflow winds couple with fast forward storm motion. This is favored by a forward-propagating storm. It has also been shown (Brooks and Doswell 1993) that some damaging events are caused not by processes within a quasi-linear line but by discrete high-precipitation supercells embedded within the line. However again it must be reiterated that the derecho is a wind event produced by a strong MCS, and is not a specific type of MCS.

5.3. Derecho Forecasting. Given the presence of sufficient instability for thunderstorms, the strongest contributing element to derecho events is significant warm air advection. A weak boundary parallel to the upper-level flow is another contributing factor. Development of MCS activity with derecho potential is most likely at the intersection of the low-level jet with this boundary.

Nevertheless, derecho events do occur across a wide range of instability and shear distributions. Some correlation has been made showing that weak mid-level storm-relative winds, strong storm-relative low-level inflow, and fast storm movement relative to the ground are all key ingredients in derecho environments. It has been found that while high CAPE values are associated with MCS events that produce high wind, they do not discriminate between severe and derecho modes of MCS systems (Cohen et al 2007). However the same study found that high values of DCAPE correlated well with derecho events,

Rise of the derecho

When I first started forecasting at the NSSFC in Kansas City in the early 1970s, all of the forecasters were told that viewing the new models and seeing how synoptic scale upper troughs and low pressure systems were moving and changing were very important since they were completely necessary for causing the development and evolution of severe convective storms. So, even though we knew we had to have enough instability and wind shear for severe weather development, we expected all severe storm events to be directly associated with the troughs that we saw on the 500 mb maps and the associated low pressure systems.

What we did not know was that occasionally severe convective storm development and evolution can be affected by storm scale features. When storms form and join together as a storm system (now called an MCS), they can sometimes affect their evolution. For example, MCS systems that develop during the warm season in areas of moderate or weak winds and in a very unstable atmosphere will sometimes move away faster than the mean winds and cause widespread severe wind damage (a derecho) well away from any synoptic scale trough. Another thing that we did not know was that an MCS system can occasionally produce a small scale low pressure system that is called a mesoscale convective vortex (MVC) and it can stay aloft well after the MCS dissipates. The MVC has rotating winds and enhanced wind shear. And if it moves into an area of sufficient instability, it can redevelop new severe convective storms that sometimes include tornadoes.

BOB JOHNS, 2009
former NSSFC (SPC) forecaster
personal communication

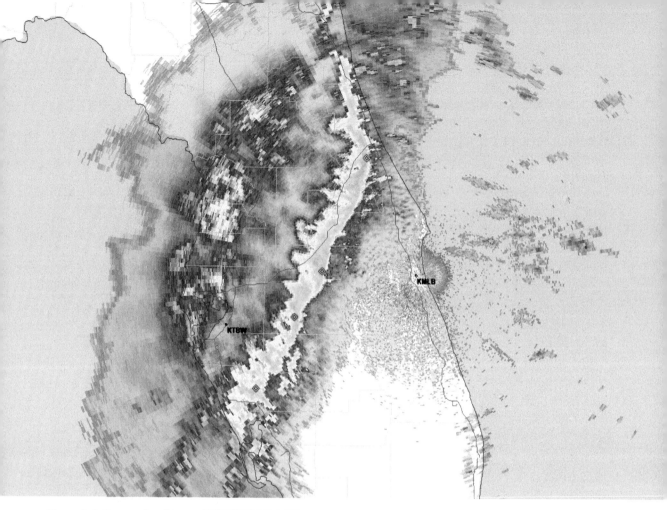

Figure 3-8. Progressive derecho. The storm of 13 March 1993 may be remembered as the Storm of the Century that brought record snow to the eastern U.S., but often forgotten is the MCS that brought massive derecho damage to the Gulf Coast states, seen here on 3/13/93 at 0617 UTC. The derecho killed 7 and injured 79. The tail end extended south to Cuba, killing 10 and causing $1 billion in damage in that country alone. Note the classic trailing stratiform precipitation area in the wake of this MCS. *(Tim Vasquez, created with GR-Level2 <grlevelx.com>)*

along with steep mid-level lapse rates and a strong decrease in theta-e with height.

6. MCS propagation

The movement of the MCS is an important forecast consideration due to the system's great potential for flash flooding. Structure, shear profiles, and boundary orientations are all important influences on MCS propagation.

6.1. MOVEMENT. A basic rule of thumb suggests that the MCS line as a whole will move with the 500 mb wind direction at 40% of this speed. Individual cells will move in the direction of the 0-6 km bulk shear vector.

An MCS flows with the mean tropospheric flow but tends to propagate towards low-level inflow. A study by Corfidi in 1996 found good results from a rule that uses the mean steering flow vector with the addition of the inverted vector of the wind at the level containing the strongest inflow. This is known as the Mesoscale Beta Element (MBE) vector, sometimes nicknamed the Corfidi vector. The steering flow vector should be consistent with the observed motion of individual cells.

It was found that the MBE predicted backbuilding MCSs well, but did poorly with forward propagating systems. The technique was adapted for forward propagating MCSs by using the MBE, as previously described, and adding that to the mean steering flow.

6.2. REAR PROPAGATING (BACKBUILDING) MCS. With this type of MCS, forward movement is negated by backward cell development, thus it is associated with slow ground-relative speeds. As a result the primary threat is flooding and "training" of cells over specific locations. Storms that have a leading stratiform precipitation area are prone to rear propagation.

6.3. FORWARD PROPAGATING MCS. Forward movement couples with forward cell development, so the result is fast ground-relative speeds. This type of storm is responsible for many damaging wind events. It is most common in northwesterly flow where the movement of the storm is directly towards the moisture inflow vector, shifting the inflow convergence to the front flank of the storm, and where a trailing stratiform mode has developed. In a severe weather regime, front-flank storms tend to be associated with non-tornadic derecho or HP-type storms.

6.4. DISCRETE INITIATION. Sometimes isolated storms will form ahead of the MCS. These are especially favored for severe weather development since competition for moisture and updraft undercutting is weak, at least in the early stages. Some current theories and modelling suggests that discrete initiation occurs beneath the downstream anvil where gravity waves associated with the anvil couple with axes of enhanced moisture left by cloud streets. ¤

The derecho

I almost regret having accepted the editors' invitation to contribute an article to the American Meteorological Journal on tornadoes . . . The real difficulty of writing on this subject consists in the fact that two radically different phenomena are described and officially catalogued under the name of tornado. Although I have years ago called attention to this fact, the practice still continued . . .

Even before organizing the first state weather service in this country, I had noticed some of the peculiarities of the storm which I now propose to call the Derecho . . . after continued study and comparison of personal observations in the field and in the observatory, as well as after charting a great many of these storms in Iowa, and considering the continued confounding of the same with the tornado, I have for a long time deemed it both wise and necessary to introduce a specific term for the truly specific phenomenon under consideration.

Since the "Twister of the Prairies" has been named the tornado, I propose to call the peculiar "Straight Blow of the Prairies" the derecho; (Spanish, in analogy with the word tornado).

GUSTAVUS HINRICHS, 1888
Iowa Weather Service director

4 TORNADOES

The whirlwind originates in the failure of an incipient hurricane to escape from its cloud: it is due to the resistance which generates the eddy, and it consists in the spiral which descends to the earth and drags with it the cloud which it cannot shake off. It moves things by its wind in the direction in which it is blowing in a straight line, and whirls round by its circular motion and forcibly snatches up whatever it meets.

— Aristotle
Meteorology, c. 350 BC

No place in the world experiences tornadoes with such frequency and intensity as the Great Plains of the United States. This region offers close proximity to cold continental air from Canada and warm, humid air from the Gulf of Mexico. The Great Plains does not have a monopoly on tornadoes, however. They have occurred in almost every temperate country in the world. Some rather substantial hot spots exist in Bangladesh and the temperate regions of inland South America. And all warm ocean regions without fail are prone to the tornado's weaker cousin: the waterspout.

Tornadoes appear within Biblical writings and the earliest theories were put forward by the Greek philosopher Aristotle in the 4th century B.C. Modern research did not begin until the late 19th century with the development of observational networks and the legendary research of U.S. Army officer John P. Finley, who established the first set of weather patterns that favor tornadoes. His findings remained a staple of Weather Bureau forecasting for nearly 50 years.

An understanding of tornadic storm structure began to appear in the 1950s. The presence of a larger "storm cyclone" surrounding the tornado was suspected in early barograph records (Brooks 1949, called a "tornado cyclone" at that time) and was linked to storm-scale rotation by Theodore Fujita. His work on the 1957 Fargo tornado identified the wall cloud and proposed a life cycle of the tornado. The supercell itself was identified as a unique storm type in 1962 by Keith Browning and Frank Ludlam, casting doubt on the longstanding notion of tornadic squall lines, and its unique structure was further confirmed in the 1960s with Doppler radar observations. By the 1970s a rather complex conceptual model of the supercell had emerged which was confirmed by research chase observations. Although meteorologists have an exceptionally good understanding of storm structure, many of the small-scale processes within the storm remain poorly understood.

Since this title is primarily an operational forecasting book, the description of the tornado provided here will focus primarily on forecasting and nowcasting aspects, with information on the internal dynamics, damage, and visual highlights abbreviated.

1887: Tornadogenesis

The sudden appearance of ominous clouds, first in the southwest and then almost immediately in the northwest or northeast (perhaps the reverse in the order of their appearance) generally attracts the attention of the most casual observer, and frequently overcomes him with astonishment. In almost all cases these premonitory clouds are unlike any ordinary and usual formation. If they are light, their appearance resembles smoke issuing from a burning building or strawstack, rolling upward in fantastic shapes to great heights. Again, like a fine mist or quite white, like fog or steam. Some persons describe these light clouds as at times apparently iridescent or glowing as if from their irregular surfaces a pale, whitish light was cast.

The dark clouds at times present a deep, greenish hue which forebodes the greatest evil and leaves one to imagine quite freely of dire possibilities. Again, they appear jet black from center to circumference, or, in a change of form, this deep-set color may only appear at the center, gradually diminishing in intensity as the outer edges of the cloud or bank of clouds are approached.

JOHN P. FINLEY, 1887
storm researcher
Tornadoes

Figure 4-1. Supercell structure was captured remarkably well in this engraving of an 1877 Pennsylvania tornado which appeared ten years later in a book by noted tornado researcher John P. Finley.

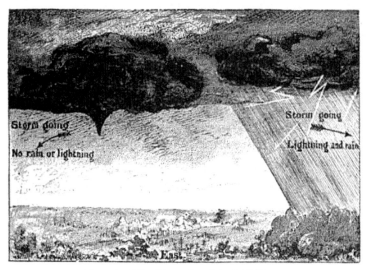

Tornado at Ercildoun, Pa., July 1, 1877.

1. Supercell tornado lifecycle

The tornado is a pendant cloud that hangs down from the base of the cumulonimbus cloud. In the taxonomy of cloud forms it is known by the Latin name *tuba*, though this name is rarely used. The development of the tornado is referred to as *tornadogenesis*.

1.1. CLASSIC LIFE CYCLE. The tornado develops from a parent cumulonimbus cloud, descending from its base or from a wall cloud or lowering attached to the base. During the initial descent stage the vortex is described visually as a funnel cloud. It should be noted that the lack of a condensation cloud connected to the ground does not imply there is no intense wind circulation at the surface, and what looks like a funnel cloud may very well be a tornado.

Once a tornado is on the ground, it produces damage which is described in more detail in later sections. It continues in this manner for anywhere from mere minutes to over half an hour. At this time a *clear slot* is often seen to wrap around the tornado from the rear side of the storm. This clear slot is an area of higher cloud which gradually turns the tornado from a cone shape into a more elongated, trunklike structure. The clear slot continues to "erode" the tornado, which tends to move to the rear of the storm.

Swirl ratio

The term "swirl ratio" is sometimes encountered in tornado literature. It is defined as the ratio between tangential (circular) and radial (inward/outward) velocity. High swirl ratio has strong rotation and weak inflow, while low swirl ratio has weak rotation and strong inflow.

Figure 4-2. Quinter, Kansas tornado on 23 May 2008 about to cross Interstate 70. This tornado was rated at EF4. Note that the debris cloud shows evidence of a toroidal structure, demonstrating that tornado flow is always made up of simple helical flow. *(Bill Hark, harkphoto.com)*

Soon the tornado begins shrinking substantially or fragmenting. This processes is informally referred to as "roping out". The tornado then vanishes from view, but significant amounts of debris may still be suspended.

1.2. VISUAL STRUCTURE. The visible tornado cloud is the result of moisture condensation within the pressure deficit zone along the tornado axis. This is the same process that causes normal cloud formation, so the funnel is essentially a cloud. Some tornadoes are manifested by a thick, well-defined condensation cloud, while others look more transparent. This is largely a function of the ambient relative humidity and the degree of pressure deficit within the funnel. However, no research has established any sort of link between the presence of a condensation cloud and tornado intensity.

The tornado picks up soil, trees, and structures, producing not only a condensation cloud but a debris cloud. In some cases such as over freshly-plowed fields, the debris cloud can become so extensive that it obscures the tornado. The 30 August 1974 Great Bend KS event remains a historically noteworthy example of such a tornado. It must be remembered that damaging winds can extend

Figure 4-3. Tornado on May 3, 1999 near Cyril, Oklahoma. Note the very bright cloud face just above and to the left of the funnel, indicative of a clear slot wrapping around the tornado. *(Tim Vasquez)*

quite a distance away from the condensation funnel and even the debris cloud. The rear-flank downdraft and even the inflow are capable of producing damaging "straight-line" winds even in proximity to the tornado.

1.3. PRECIPITATION CHARACTERISTICS. The tornado can also be obscured by rain and hail, which circulates with the tornado cyclone around it. This increases the difficulty of storm spotter operations. With storms that lean more towards the HP end of the spectrum, the precipitation can completely obscure the tornado.

Storm chasers report that precipitation near the tornado is often associated with "atomized rain", a nickname given to a spraylike fall of drizzle-sized droplets. In early government tornado surveys, victims did indeed describe a misty type of rain that followed the tornadoes. There is no clear explanation what causes this phenomena, and not all tornadoes are associated with it. Likewise, very large drops are sometimes found around the tornadic circulation and are attributed more to hailstones that have completely melted during their fall to earth.

1.4. CLIMATOLOGY. Tornadoes in the United States are most common during the month of May, with records showing a diurnal peak of 5 p.m. local standard time. Peak occurrences, however, vary strongly by region. For example in Canada and the northeast United States, tornado occurrence is most common during the middle of summer, while along the Gulf Coast, the winter months bring the greatest threat. Though Oklahoma City and regions west and southwest are identified on many maps as the world's hot spot for tornadoes, some new research work done at NSSL may shift this axis to western Kansas, the eastern Texas Panhandle, and adjoining sections of western Oklahoma.

2. Supercellular tornado mechanisms

The workings of tornado formation have been the subject of idle speculation for centuries, but with expansion of populated areas into tornado-prone territory, this has become a very serious rescarch topic involving millions of dollars in research money. Challenges include the difficulty of obtaining meaningful measurements of wind, temperature, and moisture throughout the storm volume as well as the complexity of fluid dynamics and its interactions with the storm and the ground. Storm modelling has helped, but some of the greatest advances in recent years have come from field experiments such as VORTEX.

A century ago, tornadogenesis was believed to originate from conservation of angular momentum. As inflow arrives into the supercell, the horizontal convergence of mass must be compensated by increasing vertical vorticity, giving the storm spin. To some extent this accounts for the large-scale mesocyclone but falls far short of explaining the tornado. The growing complexity of the supercell model has led to a host of other theories, including bizarre ideas about electrical and magnetic energy which were often floated in journals during the 20th century.

2.1. MESOCYCLONE FORMATION. Since the 1960s, research has shown a strong link with horizontal shear in the pre-existing environment between the surface and the cloud layer, in other words, a significant difference in winds with height. It is thought that

---

Tornado theory 1899

Accounts of tornadoes and waterspouts frequently mention the descent of their funnels from the heavy overhanging clouds, as if there were actually some descending motion; but there is good reason for believing that the descent is only a deceptive appearance. While it is generally true that the movement of clouds indicates the movement of air in which they are formed, this is not always the case ... The funnel seems to descend, because, as Franklin clearly said in 1753, the moisture is condensed "faster in a right line downward than the vapors [cloud particles] themselves can climb in a spiral line approach."

WILLIAM M. DAVIS, 1899
Harvard University researcher

> **A 1914 tornado theory**
>
> The oppositely directed adjacent, not, conflicting upper currents by catching masses of air, especially by rising masses, between them tend mechanically to produce, in the middle atmosphere, violent vortices of limited extent. The more violent of these vertical atmospheric whirls, usually accompanied by thunder and rain and often extending down to the surface of the earth, where they become destructive, are known as tornadoes.
>
> Hence thunderstorms generated in the barometric region under discussion [between the branches of a V-shaped isobar, i.e. a trough], the region in which tornadoes most frequently originate and develop, might properly be called "tornadic thunderstorms"
>
> W. J. HUMPHREYS, 1914
> Weather Bureau meteorologist

this shear produces horizontal vortices whose axes are aligned with the low level winds, yielding streamwise vorticity. This air is then drawn up into a storm, tilting the vorticity into the vertical and providing a source of horizontal vorticity in the storm cloud, enough to produce a mesocyclone. This is an area of rotation in the cloud measuring on the order of miles in diameter.

Stretching of vorticity into the vertical is an effect that increases the rotational speed of the vortex. This occurs when wind along the vorticity axis is diverging, in other words, when air at one end is moving away relative to the other end. This effect is most pronounced where the vorticity axis extends up into the updraft where air parcels are accelerating. Together, low-level environmental shear and stretching produces the large mesocyclone seen in a parent supercell.

The mesocyclone, however, is unable to provide the rotation needed for tornadogenesis. Furthermore if the tornado was merely produced by a mesocyclone it would tend to develop in the middle of the rain free base and would frequently occur with any strong supercell, even an LP storm, which is known for its strong mesocyclones but its lack of tornadoes.

2.2. THE DOWNDRAFT. The role of the downdraft has been solidly established in tornadogenesis. During the 1970s Theodore Fujita found evidence of divergent microburst damage near the beginning of a tornado track. This suggested the existence of a separate downdraft on the rear side of the supercell, which is now known as the rear flank downdraft (RFD). Research during the 1980s found strong correlations with rear flank downdraft (RFD) characteristics and tornadogenesis. Likewise, a growing body of field evidence suggested that LP supercells were poor tornado producers and rarely produced a significant RFD at the surface. This distinction within the supercell spectrum helped point the way toward the role of the downdraft.

Reviewing 1990s-era VORTEX studies, Markowski et al. 2002 found that the character of virtual temperature and equivalent potential temperature (theta-e) in the RFD region of the storm has an association with the mode of tornado formation. A cold RFD, which has temperatures significantly colder than the inflow air at the same level, appears to interfere with tornadogenesis. A warm

RFD, which has temperatures nearly equal to or slightly cooler than the inflow, favors tornadogenesis. In a warm RFD, it is not as negatively buoyant and its tendency to produce outflowish downdrafts are minimized. A warm RFD may be caused by entrainment of updraft air that has not risen very far, or in situations where poorly mixed air reaches the surface. Weak subsidence of a warm RFD may also promote shallow trajectories which enter into and shape the tornadic circulation.

What this implies for the forecaster is that tornadogenesis is most likely in storms whose rear-flank downdrafts are atypically warm. In most cases this temperature is less than 5 C° cooler than the inflow air or even nearly the same temperature. To achieve this, it appears that a moist air mass is necessary, and underscores the importance of low LCLs.

2.3. SPINUP MECHANISMS. Once a downdraft is established and is in place, how does the tornado actually form? Unfortunately this is where the limits of our knowledge are strained, replaced with conjecture, as the spinup mechanisms are on the microscale and generally escape detection. There is agreement, however, that strong vertical accelerations within the thunderstorm contribute *stretching*. Mass conservation dictates that when a rotating axis is stretched, it must rotate faster to preserve its angular momentum. In a thunderstorm, an updraft or downdraft all provide mechanisms which act to stretch a vertical vortex. One theory speculates that vorticity along the "skin" of the descending RFD breaks down into vortices which then undergo stretching, similar to the stretching of the pre-existing shear environment mentioned earlier.

It is thought that friction near the ground at this stage may augment spinup somehow by slowing air near the ground just outside of the tornado, allowing air to flow more directly toward the vortex. This concentration of mass near the axis dramatically increases rotation due to conservation of angular momentum. Normally this influx of air fills the void, but it is suspected that in some cases the vortex is vented by a very strong updraft. The strong updraft removes the mass that constantly enters the system and prolongs the life of the rapidly spinning vortex.

Once established, the vortex builds downward to the ground through the *dynamic pipe effect* (DPE). It effectively creates a con-

---

Small scale vorticity

Common sources of vorticity at the ground include rotation within microscale low pressure areas, rotation along gust fronts, and even rotation along cloud shadows. Although these very small-scale vorticity circulations are often microscale in character, they are sometimes referred to as misocyclones, after the 1981 Fujita classification which defined 40 meters to 4 km as the "misoscale".

F-scale stats

Tornado track width and length as related to Fujita damage intensity, as indicated by Brooks (2004). Based on a random 4-year period.

F-scale	Mean length (km)	Mean width (m)
F0	2	30
F1	5	70
F2	10	120
F3	22	270
F4	43	460

Overpass dangers

Since the late 1990s, tornado damage experts have been working to steer the public away from a particularly dangerous form of shelter: the enclosed spaces underneath highway overpasses.

* It is nearly impossible to hide the body from the violent airstream underneath the bridge, which is full of glass, splinters, dirt, and other debris.

* Many bridges have shallow or nonexistent recessed space. When people find this out it's often too late.

* Air conserves energy by accelerating through confined spaces and around barriers.

* Popular media clips of overpass strikes (e.g. 1991 in Andover) which are frequently aired on cable TV programs actually take place at the outer periphery of the tornado circulation.

* On 3 May 1999, three were killed hiding under bridges.

Figure 4-4. Actual warning operations on 15 May 2003 at the Fort Worth NWS office. In this frame, warning coordination meteorologist Gary Woodall monitors a tornadic storm northwest of Abilene using storm-relative velocity products. *(Tim Vasquez)*

stricted vortex in which air cannot enter the sides due to strong centrifugal force, but may enter from the bottom where the vortex has not spun up and centrifugal force is weak. The air entering the lower portion of the vortex spins up and effectively becomes part of the vortex itself. This builds the vortex downward. Air may no longer enter this level and must flow in from below.

It is generally believed that some tornadoes, primarily the typical supercell tornadoes and cold-air funnels, grow mainly from the dynamic pipe effect. In other types of tornadoes, such as landspouts, gustnadoes, and mesovortices, the growth mechanism is limited primarily to vertical stretching of pre-existing vortices. There is also some thought that cyclic supercell tornadoes which occur after the initial occlusion may originate mainly from stretching associated with significant low-level forcing.

2.4. TORNADOGENESIS FAILURE. A supercell may fail to develop a tornado, even when formation seems imminent and a wall cloud is rapidly rotating. The causes of failure are just as puzzling as the causes of formation. The properties of the RFD downdraft do appear to play some part in tornadogenesis failure. This may occur from an excessively cold RFD or from prodigious outflow that

dominates the storm or somehow overpowers the spinup mechanisms. One of the reasons that tornadoes are not common in HP supercells is because of the strong RFD production.

2.5. TORNADO VORTEX STRUCTURE. As the rotational speed of the tornado increases, a pressure deficit forms along the axis of rotation, in other words, within the core of the tornado. This pressure deficit area contains poor swirl ratio; in other words, the air inside of it is relatively calm. The deficit area tends to descend inside the core of the tornado like a downdraft along the axis toward the ground, producing what is called axial downflow. The axial downflow's interface with the surrounding, rapidly-rotating air is called the breakdown bubble.

As the breakdown bubble approaches the ground, the tornado overall is highly developed and has usually reached its strongest wind speed values at the surface. Once the breakdown bubble reaches the ground, the laminar structure of the vortex is destroyed and vortex breakdown occurs. The tornado becomes a multiple-vortex tornado composed of smaller vortices, which may or may not be seen. These individual vortices can be very strong and produce localized, enhanced damage. During this change, an occlusion downdraft usually begins affecting the tornado.

During the 1970s and 1980s, it was often held that weak tornadoes were comprised of a single vortex, while violent tornadoes broke up into multivortex structures. This has been found to be incorrect. Most tornadoes tend to have all of these characteristics at some point during their life cycle, and the anomalous structures are simply more evident and easier to see in the stronger tornadoes.

2.6. OCCLUSION DOWNDRAFT. During the tornado's life cycle, an occlusion downdraft, embedded in the rear-flank downdraft, develops behind the tornado (relative to the mean tropospheric flow) and works its way around the tornado. This is manifested by a higher cloud base or vaulted area (a "clear slot") or a rain curtain which gradually wrap around the vortex.

2.7. CYCLIC TORNADOGENESIS. During the 1960s it was observed that supercells produced not just one tornado but many others, sequentially, and at regular intervals. This characteristic has come to

Infrasonics

There is some evidence that severe thunderstorms emit distinctive acoustic signatures, which are in the infrasonic range. This is at about 0.1 to 10 Hz, well below the threshold of human hearing. A review of peer-reviewed journals indicates that little in the way of institutional research has been done in this area, however some preliminary work published in 2008 suggests the sounds may originate from in or near hail cores. Researchers are hoping to developing a numerical model of severe thunderstorms adapted for acoustic simulations, however the main limitation is that grid spacing on the order of 1 meter is needed, requiring extreme computational power. Infrasonic tornado detection remains an embryonic area of study and its suitability for forecast use will not be known for some time.

Figure 4-5. Cyclic supercell near Stratford, Texas in May 2003. The view is looking north. The older tornado can be seen to the left, with new development taking place center foreground. *(Olivier Staiger)*

be known as *cyclic tornadogenesis.* The traditional understanding of cyclic tornadogenesis is that outflow and precipitation eventually wrap around the mesocyclone and tornado, occluding it much like the occlusion of a frontal system. The old tornado moves into the rear of the supercell and dissipates, while a new mesocyclone and tornado forms further ahead where the updraft and inflow are unaffected.

3. QLCS tornadoes

Tornadoes may also originate from quasi-linear convective systems: that is, mesoscale convective systems with linear characteristics where deep, persistent mesocyclones can be ruled out. These involve derechos, bow echoes, book-end vortices, cold pools, and rear-inflow jets. Although damaging straight-line winds are the most common threat from these systems, brief tornadoes can occur and they may be difficult to detect. The term QLCS was introduced to differentiate supercell tornadoes and landspouts from those in linear MCS storms.

3.1. BOW ECHOES. Even though a QLCS may have a two-dimensional structure, irregularities often occur along gust fronts. These produce varying amounts of horizontal shear and may amplify into *mesovortices* measuring hundreds of meters to kilometers in size. The accelerating bow echo produces a zone of cyclonic shear that extends from the apex poleward (northward in the northern hemisphere) and this is the favored area for mesovortex generation.

This process, under certain conditions, can result in tornadogenesis along the leading edge of the MCS. It is thought that the strengthening rear-inflow jet within a bow echo may generate horizontal vorticity that can tilt into the vertical. The horizontal vorticity source is the shear between the storm's cold pool and the environment, not the pre-existing synoptic-scale or mesoscale shear. Vertical stretching then intensifies the tornado.

Numerical modelling also shows that upright updrafts are more likely to promote intense mesovortices, so a bulk shear that is neither too strong nor too weak may favor this type of mode. Orthogonal intersections of MCS systems and bow echoes with pre-existing boundaries have been shown to correlate with tornado development.

3.2. BOOKEND VORTICES. A bow echo produces a forward thrust of wind. This induces cyclonic rotation on one end and anticyclonic rotation on the other, much like the turning of the hinges when one walks through a double-door. Each end is referred to as a line-end or bookend vortex (Weisman 1993). The poleward end takes on cyclonic flow and the other end anticyclonic. A tornado may occur in this area on the rear side of the echo and are frequently embedded in rain.

In general, bow echoes are not significant tornado producers and it is not well-understood what additional prerequisites are required for tornado formation. Furthermore, the idea of the bow echo head simply being a supercell with a deep, persistent mesocyclone is not widely accepted.

4. Miscellaneous tornado types

There are also a number of tornado types which can result from highly localized processes or from horizontal shear. These can

---

Waterspout legend & lore

Waterspouts often leave behind a sulphurous smell, and there are examples of a disagreeable smell remaining along the whole tract traversed by them. One individual, however, who became involved in a waterspout perceived no odor.

We seldom read accounts of waterspouts without finding also that electrical phenomenon were noticed at the same time. Lightning is almost never wanting; thunder is likewise often connected with them, and it has been remarked that the loud noise which follows waterspouts easily prevents feeble peals of thunder from being heard.

Now and then, a more widely dispersed light has been seen; so that people imagined that the corn in their fields was on fire, but afterwards to their joyful astonishment found it uninjured. It has been reported of one waterspout that fireballs proceeded from it, one of which was accompanied by a report like that of a musket.

HANS CHRISTIAN ØRSTED, 1839
Danish physicist

occur in almost any type of storm, including in supercells with classic tornadoes.

4.1. WATERSPOUTS. Over warm oceans, thunderstorms are quite capable of producing an equivalent phenomenon: the classic waterspout. The waterspout begins with a funnel cloud and a disturbance on the ocean surface which then produces a more well-developed condensation cloud and a spiral pattern on the water's surface. The funnel soon decays and dissipates.

4.2. LANDSPOUTS. For many years it was recognized that waterspouts had a benign character and were often associated with nonsevere storms. However in 1927, British researcher H. G. Busk documented benign tornadoes with a smooth, laminar vortex in Persia and gave them the name of "landspout". In the 1980s storm chasers noted waterspout-like tornadoes on the Colorado plains descending from nonsupercellular storms. These Colorado observations brought the term into the lexicon of American forecasting. Their unique character was underscored in a 1980s radar study by Ray Brady and Ed Szoke which found no mesocyclone with these events. It is generally accepted that landspouts and waterspouts occur due to the stretching of pre-existing vertical vorticity, such as that in small-scale heat lows, or from the tilting of horizontal vorticity along boundaries by a strong updraft. Landspouts are capable of producing damage but are normally very small and tend to move slowly.

4.3. GUSTNADOES ("gustinadoes" in 1980s literature) describes a brief spinup near the ground along an outflow boundary. They usually have a scale on the order of seconds and tens of feet and are most common along gust fronts in and around the storm, including in squall lines. A gustnado may briefly look like a small tornado without connection to the cloud base. There is no correlation to a mesocyclone.

Since they occur on such a small scale, they cannot be predicted, thus in the United States they are recognized in warnings with the blanket statement that "storms can and do produce tornadoes with little or no advance warning". In spite of their size, gustnadoes are capable of producing F0 to F1 damage.

4.4. DUST DEVILS. Dust devils generally occur under clear skies with intense heating, though they can be favored along the edges of cumuliform cloud shadows. The circulation can reach to many thousands of feet with life cycles of minutes, and damage of EF0 to EF1 can occur with the strongest events. Dust devils are not operationally forecasted and have received little academic and research attention.

4.5. COLD AIR FUNNELS. The so-called "cold air funnels" are not even tornadoes but are very small funnels that occur on the edges a cumuliform cloud. They occur on a scale of mere seconds and tens of meters. As with landspouts, they are formed from the stretching of vertical vorticity near the cloud. The name misleadingly suggests a causal relationship with cold air aloft, but it is recognized that the funnels are likely associated with relatively benign cloud-scale shear and vorticity processes. They are not thought to correlate with any damage at the surface and are not considered to be significant to forecasting.

5. Tornado damage

While differentiation of a tornado event from a microburst event is made easy by photographic evidence, in many cases no such evidence exists. Damage surveyors analyze the debris field along the damage axis to deduce wind direction, and search for convergent or divergent characteristics. Deducing rotation by simply looking for opposing wind direction along the axis is not reliable, because a fast-moving tornado produces light damage along one side of its track and significant damage along the other side. This can produce gross errors in placing a tornado centerline as all damage found across the axis may appear to be strewn in the same eastward direction.

Damage tends to be stronger on the side of the tornado track where translational (tornado movement) speed couples with rotational speed. In the northern hemisphere, this is usually on the south side of the track. This effect is particularly pronounced with fast-moving tornadoes, such as those that occur during the cool season. For example, given a typical Kansas tornado with tangen-

The devil in dust devils

When I was young, I figured that dust devils were nothing but little whirls that might be seen out in the desert, tossing up a few tumbleweeds. That changed for me April 14, 1984, in Tucson, Arizona. Representing the drafting class, I managed a small vendor table at an outdoor swap meet (flea market), where our high school was selling donated goods to raise money for the school.

It was a completely clear day with temperatures in the 90s, and business was slow. I was just starting to read a book when I heard people yelling. I turned around and witnessed a huge elephant trunk of dust, at least 1000 ft high, crossing the parking lot and coming for us.

As dust, weeds, and debris started whipping past us, instinct took over and I ran behind the open door of a pickup truck parked nearby. As the wind subsided, I saw the dust devil had become infiltrated with garments, papers, and merchandise. I looked around for the videogames we had been selling, but found no trace of them. Many tables had been swept clean. Right then the swap meet turned into a free-for-all treasure hunt with dozens of people fanning out looking for cash. Some folks followed the path of the dust devil into the neighborhood looking for riches. Needless to say, the swap meet closed for the day.

TIM VASQUEZ

Tornado wind speeds

Early in the 20th century, tornado wind speeds were believed to range as high as 400 to 800 mph. These beliefs began to change following the photogrammetric work of Walter Hoecker, who analyzed the 1957 Dallas tornado and found winds of 170 mph.

Figure 4-6. Devastating F5 tornado damage at Bridge Creek, Oklahoma from a supercell that tracked from Lawton to Oklahoma City on 3 May 1999. *(Tim Vasquez)*

tial wind speeds of 100 mph moving at 25 mph, the left side of the track receives 75 mph winds with the right side receiving 125 mph winds.

Though there has been loose association with tornado size and strength, it has been determined that there is in fact a close correlation with track width and track length and rated intensity (Brooks 2004). It should be emphasized that visual size of the condensation or debris cloud does not provide a good estimate of the actual damage path width.

In stronger tornadoes, the damage path often shows cycloidal tracks on the scale of tens of feet. So-called "suction spots" are also sometimes found, which may be associated with vortex breakdown of the primary funnel into multiple vortices. A linear deposition zone of debris may also be found directly along the tornado track. This deposition zone is attributed to boundary layer inflow into the rear of the tornado.

5.1. FUJITA SCALE. The original Fujita scale was developed in 1971 by Theodore Fujita and was used until 2007. The scale is a scientific attempt to estimate tornado wind speeds by evaluating the tornado's damage to residential houses. The tornado then receives the designation corresponding to the strongest damage found. From its outset, the scale was intended to have an arbitrary relationship to wind speed. It is based on the equation $V = 14.1(F+2)^{3/2}$ where V is the velocity in mph and F is the F-scale number. Many difficulties in estimating damage have come from failures due to impacts from other flying debris, inadequate sampling due to a "near miss" from a tornado, and a wide range of construction quality.

The Fujita scale has often been misused to visually describe a tornado. Tornado intensity cannot be reliably estimated by sight. The idea of an F6 tornado has also come up from time to time. Though Fujita's extended scale does allow for it, the damage that would indicate an F6 tornado is by definition "inconceivable". For this reason, there has never been any tornado designated as an F6.

5.2. ENHANCED FUJITA (EF) SCALE. During the 1990s it was recognized that the existing Fujita scale hinged upon the tornado's effects on specific types of well-built structures. Texas Tech University, a leader in storm damage engineering, recommended in 2004 the adoption of a rating system with more indicators, more flexibility, and better correlation with actual wind speeds. The result was the Enhanced Fujita scale (see Appendix). It was adopted by the United States on 1 February 2007. The Enhanced Fujita is similar to the Fujita scale, ranking tornado damage on a scale from EF0 to EF5. Engineers performing the ratings evaluate 28 damage indicators (DI) for various types of construction, for which one of 10 degrees of damage (DOD) are determined.

5.3. SAFETY. As of 2010, current tornado safety philosophy emphasizes the importance of getting to the innermost room on the lowest floor of a well-constructed building. Driving is a major danger due to the possibility of traffic jams, toppling high-tension power lines, and deteriorating visibility, but if the tornado is still at least several miles away, motorists may turn onto non-congested roads that are at right angles to the tornado's path and which lead away from the storm. If the tornado gets close or cannot be seen, swiftly take shelter in any strong building. Any well-constructed building is far safer than in a vehicle. Never hide under an overpass.

Many unsubstantiated myths flourished in the 20th century regarding tornado safety and some are still heard today. Opening garage doors, opening windows, using the southwest corner of a basement, and hiding underneath an overpass have all been shown to be ineffective and in some cases can increase the danger.

6. Tornado prediction

There are three primary requirements for significant tornadoes that must coexist: a persistent updraft, special rear-flank downdraft characteristics, and enhanced storm-relative helicity. Also a

---

The pride of man

When the author first advised the building of tornado caves [underground shelters] in 1879 as a means of saving life people were incredulous and the Press, in some instances, rather ridiculed the idea saying, that at the least, it was rather an undignified thing for man, the grandest and greatest of all creation, to burrow in the ground at the sight of a dark cloud.

JOHN P. FINLEY, 1888
Tornadoes, What They Are, and How To Escape Them

The falling fatality rate

In a 1998 paper Thomas Grazulis noted that the population-adjusted tornado fatality rate began a strong decline after 1925, which he ascribed partly due to public awareness following the 1925 Tri-State tornado disaster, societal changes as rural populations declined during the Great Depression, and the advent of radio and television. The 1950s shift of the U.S. Weather Bureau from its reserved stance on storm forecasting was also a factor.

Post WWII forecasting

Empirical tornado forecasts began in 1948 with the work of Miller and Fawbush. The Weather Bureau began to investigate dynamic ingredients, starting with the intersection of pressure jump lines (Tepper 1950) which were suspected to synergistically increase convergence and vertical vorticity. The work of the Severe Local Storms Unit in the 1950s dramatically improved the body of knowledge for tornado forecasting.

low LCL is important, as otherwise high-based storms are implied, which are not generally associated with tornado production. Assuming the environment is supportive of long-lived storms, the following list summarizes some of the key issues which were considered significant to tornado forecasting in 2009.

6.1. LOCALIZED ENHANCEMENT OF SHEAR. Baroclinic generation of vorticity occurs wherever temperature contrasts occur. The wind is locally backed within a baroclinic zone, and this pattern can persist even after temperature gradients diminish. Locally backed winds promote greater amounts of storm relative helicity and streamwise vorticity. Mesoscale baroclinically-generated vorticity, such as that along outflow boundaries, is an important and very common source of enhanced shear for tornadic storms.

Gamma-mesoscale (scale of 2-20 km) enhancement of shear, such as along the gust front of the forward flank downdraft, has long been thought as a mechanism for tornadogenesis, but research suggests that this enhancement is on too small of a scale to be significant. Parcel residence times are only on the order of 10 to 30 minutes. However, enhancement along the edge of an anvil shadow ahead of an updraft is thought to be large enough and significant enough to augment tornadogenesis in some cases, allowing parcel residence times on the order of a couple of hours in this zone. The thermal contrast across the shadow edge is often strengthened by evaporative and radiative cooling underneath the anvil.

A boundary with high theta-e (equivalent potential temperature) air just north of the boundary is the best candidate for tornadic storms (Markowski 2002). This is colloquially referred to as "pooling of moisture", which is misleading as there is no actual "ponding" of moisture taking place. Often this zone is only about 40 miles in width. On surface charts, this can look like "pooling" of moisture just north of the boundary.

6.2. WARM RFD. The existence of a rear-flank downdraft which is not cold (that is, not excessively cold relative to the inflow) favors tornadogenesis. Specifically, an equivalent potential temperature that is not significantly lower than that of the updraft is important. It is not possible to for a forecaster to get RFD measurements in real-time, so there are certain indicators that must favor weak downdrafts. High relative humidities will favor downdrafts that have not entrained dry air, an ingredient which enhances evaporative cooling and adds to downdraft density. Examples of high relative humidity indicators in the subcloud layer are low surface dewpoint depressions, low cumulus bases (less than 3000 to 5000 ft), and low condensation levels (less than 1000 to 1400 m). High relative humidity often occurs with tornado failures, but low relative humidity tends to preclude tornadogenesis.

6.3. HELICITY. A high amount of 0-1 km AGL storm-relative helicity is considered to be an excellent indicator of tornado risk. Modified hodographs should be examined for curvature in this layer and evaluated against forecast storm motion. Boundaries which may locally increase the low-level shear should be located and identified using surface, satellite, radar, profiler, and mesonet data. Furthermore, storm rotation can come not only from vertical shear but from the horizontal vorticity present along shallow boundaries. In many storm situations, persistent boundaries mark the difference between tornadic and nontornadic modes.

6.4. ANVIL LEVEL WINDS. Moderate to strong anvil level winds appear to be required for a significant tornadic storm. This ventilates the

---

Why a warm RFD? Why a cold RFD?

In all types of convective storms, downdraft and outflow temperatures are controlled by the amount of evaporation, melting, and sublimation of hydrometeors. As a result, downdraft/outflow temperatures are influenced by the relative humidity of both the subcloud layer, the relative humidity of the environmental air that is entrained into a storm aloft, the amount of entrainment, the microphysical properties of the hydrometeors (the amount of latent cooling for a given mass of hydrometeors is sensitive to the surface area-to-mass ratio of the hydrometeors, e.g., for a given mass of liquid water, less evaporative cooling occurs if the liquid water is distributed among a smaller number of large raindrops than a larger number of small raindrops), the liquid water content of the updraft (a strong function of CAPE), and the height of the melting level.

Obviously, a number of these things are difficult or impossible to assess in real-time, let alone in field experiments. Thus, it is a challenge to forecast the strength of convective outflows with great precision for any type of convective storm.

Although all of the above effects play some role in dictating the outflow properties, in studies of supercell outflow (both within the rear and forward flanks of the storm) it turns out that the boundary layer relative humidity of the inflow, even if measured only at the surface (e.g., by examining the inflow dewpoint depression), is not a bad "zeroth-order" predictor of outflow strength. The surface dewpoint depression in the inflow accounts for roughly 30-60% of the variance of outflow temperatures, with outflow temperature increasing with decreasing dewpoint depressions (i.e., humid boundary layers suppress cold pools).

Although boundary layer relative humidity is hardly a perfect predictor of outflow temperatures, we are extremely fortunate that outflow temperatures are as highly correlated with boundary layer relative humidity as studies show, for boundary layer relative humidity is easy to observe (much easier than the liquid water content or dropsize distribution within a storm!). And given what we now know about the relationship between outflow strength and tornadogenesis (tornado likelihood decreases as outflow becomes colder), having even a crude way to anticipate outflow strength is very valuable to forecasters attempting to discriminate between tornadic and nontornadic supercell environments.

Studies from several independent investigators have shown that low LCL heights (small dewpoint depressions) derived from inflow observations favor tornadic supercells, all else being equal. It's my opinion that low LCLs are favorable because outflow tends to be suppressed in such environments.

PAUL MARKOWSKI, 2009
personal communication

storm and allows the forward flank downdraft cascade to descend further downshear. Too much upper-level wind, however, may result in highly sheared cumulonimbus clouds and low-precipitation supercell modes.

6.5. CLUSTER AND CORRIDOR OUTBREAKS. Early on, tornado groupings on a particular day were known to have various characteristics. These were resolved into local, clustered within an area about 100 x 100 miles; progressive, advancing generally west to east and line outbreaks with tornadoes focused along a line hundreds or thousands of miles long (Galway 1977). Moller (1979) noted that Texas tornado outbreaks tended to organize into either clusters or corridors. Corridors were most common in the early and mid-spring, with clusters more common in late spring. Winds at 500 mb with cluster outbreaks averaged 30 kt, while with corridor outbreaks they averaged 45 kt.

7. Tornado nowcasting

Tornado nowcasting is an issue of detection, which in itself is very challenging even with all the right tools. As of 2010, tornado nowcasting resources were limited primarily to radar, spotter networks, media outlets, and direct visual observation if available.

7.1. RADAR. Since radar is such an important tool in tornado nowcasting and aspects of severe storm structure must be considered, this topic will be discussed in depth in the radar chapter.

7.2. SPOTTERS. The role of trained spotters cannot be understated, since radar detection techniques have limitations and effective radar coverage is not available everywhere. For decades in the United States spotters have been used as the eyes and ears of the National Weather Service as a network called SKYWARN. Canada operates a similar network called CANWARN. Spotters are trained amateur radio operators who are positioned near and within the storm, monitoring and reporting storm character, rotation, and trends.

Tornado warnings in 1899

The great difficulties in the way of sending a warning forward to the next town are three.

First, you do not know exactly which way the tornado will move as a whole, and you may warn the wrong town; the present storm is said to have moved at first toward the northwest and then to the southeast.

Second, the tornado frequently retires to the clouds and is no longer felt on the earth.

Third, everyone, even the telegraph operator, is busy looking after his own safety, and when the word comes, "look out for the tornado," scarcely any one has the self-sacrifice or the self-possession requisite to call up "central" and spend several minutes sending off the necessary dispatch to the next town. Once or twice it has happened that the telegraph operator has sent the word "tornado" on to the next station but this can not be expected to happen, as a rule, in ordinary small country telegraph and telephone offices.

U.S. Weather Bureau, 1899

Even forecasters who are not part of the warning process will benefit from using a VHF scanner to monitor local spotter networks.

7.3. STREAMING MEDIA. The advent of mobile Internet has made live dissemination of photos and video from the field a reality. This data is easiest to view on local television networks which have an emphasis on severe weather. Some of the data can also be viewed directly on the Internet from participating chasers. This ability of course is subject to outages since the technology is dependent on functioning cellular networks. Even when mobile Internet fails, television news crews and helicopters may be active in the field and linked to the studio via microwave.

7.4. VISUAL NOWCASTING. This section will not summarize visual nowcasting techniques, since most of the essentials of storm evolution and structure have already been covered in previous sections and chapters. Techniques are based on application of these principles. ¤

5

Hail

Potter, Nebraska: The hailstones were simply great chunks of ice, many of them three and four inches in diameter, and of all shapes — squares, cones, cubes, etc. The first stone that struck the [Union Pacific] train broke a window, and the flying glass severely injured a lady on the face, making a deep cut. Five minutes afterward there was not a whole light of glass on the south side of the train, the whole length of it. The windows in the Pullman cars were made of French plate, three eighths of an inch thick, and double. The hail broke both thicknesses, and tore the curtains to shreds.

— Nebraska Hailstorm, 1875
Scientific American

Hail is by far the most damaging form of precipitation. The insurance industry reports hail claims in the United States total up to $2 billion per year. A single hailstorm that moved through central Missouri on 10 April 2001 produced $1.5 billion in insurance claims. In 1995, hailstorms that hit DFW Airport temporarily grounded part of the fleet, costing American Airlines millions of dollars in lost flight schedules and leading Wall Street to warn investors of dampened quarterly earnings.

In the United States, large hail is considered to be any hailstone that equals ¾ inch in diameter, e.g. dime-sized, or larger. Very large hail is considered to be 2 inches in diameter or larger and gets special attention on Storm Prediction Center graphics. For the most part only supercells are capable of producing very large hail.

1. Hail formation

The requirements for large hail are fourfold: an intense updraft, a large area of supercooled water in the updraft, long residence times of hail embryos in this updraft, and slow melting of the hailstone as it falls. Unfortunately research has been slow to form an understanding of how these ingredients interact, and even the subject of the hailstone trajectory is a subject of much speculation.

1.1. HAILSTONE FORMATION. The mid-level regions of a thunderstorm, that area between 0 and -40 deg C, are supportive of *multiple forms* of water: supercooled water, ice crystals, ice crystal aggregates, graupel, rain drops, and even ice pellets. All of these particles are candidates for becoming a hail embryo — in particular, ice crystal aggregates and ice pellets are favored.

When a very strong updraft is occurring, this generates a very large mass of supercooled water through the 0 to -40 deg C region. When a supercooled water droplet collides with any of the above ice forms, it immediately freezes. This produces a rimed particle such as graupel or a rimed ice aggregate.

Now that we have a rime ice particle, how do we make it grow? An embryo can achieve *wet growth* with a coating of clear ice. This is caused by an encounter with a particle with relatively warm temperature or large thermal mass, usually a large supercooled drop or a supercooled droplet warmer than -25 deg C. The latent heat

---

A Texas hailstorm

This remarkable hail fell in large lumps, ranging from three to six inches in diameter. I heard of one piece eight inches in diameter which weighed four pounds. They were, as a rule, spherical in form, but some were somewhat flat, and nearly all were covered with oval knobs. They fell in small areas about two feet apart, while in other places only one would fall in a space twenty feet square. The average under my observation was about one hailstone to every three feet square. A most remarkable fact in connection with these large hailstones is that some of them have particles of dirt in the center.

author unknown
account of a hailstorm in Gray Hill, Texas, in 1892

Know your particles!

Nuclei: A phase change from water vapor to a liquid or solid normally cannot occur without a hygroscopic particle to cling to, such as salt, dust, pollutants, or other lithometeors. Without nuclei the air will tend to become supersaturated.

Liquid water: Cloud drops and rain drops at temperatures above 0°C are considered to be liquid water.

Supercooled water: Pure water droplets moving into or forming within a layer of subfreezing temperatures are considered "supercooled". At 0°C to -20°C, air saturation preferentially produces supercooled droplets rather than ice because deposition is dependent on ice nuclei, which are rare, and spontaneous deposition without ice nuclei does not occur.

Graupel: When supercooled water droplets freeze after colliding with ice crystals, it causes riming onto the crystal and forms graupel, tiny sleetlike particles less than 1 mm in diameter.

Ice crystals: When water vapor undergoes deposition onto an ice nuclei, it produces crystalline ice. Cirrostratus clouds are comprised largely of these. Ice nuclei are rare, making heterogeneous deposition insignificant, but freezing of supercooled droplets may produce microscopic splinters that act as ice nuclei.

Ice crystal aggregates: Ice crystals which have collided form aggregates, also known as snowflakes. Cirrus clouds are made up of ice crystals and ice crystal aggregates.

Ice pellets (sleet): Raindrops which have frozen, and tend to be translucent rather than opaque. Sleet at the surface almost always involves a winter weather pattern.

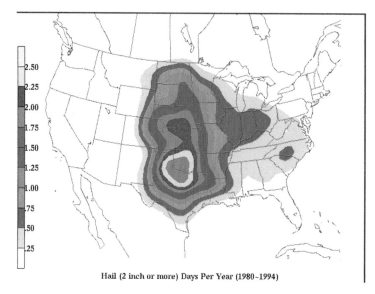

Figure 5-1. Large hail incidence in the United States for hailstones 2 inches or more in diameter. *(NSSL)*

release from freezing briefly exceeds 0 deg C and allows the outer coating to flow in a liquid form, filling all of the pores of the embryo evenly. All the microscopic air bubbles have a chance to leave the liquid coating. Freezing eventually occurs due to exposure of the layer to cold air and the ice underneath. Once this is completed, the particle has a coating of clear, transparent ice. This process of wet growth is highly conducive to further growth, as other ice particles stick readily to the particle during the brief period when it is wet rather than bounce off it.

The hailstone can also achieve *dry growth* with a coating of rime ice. This takes particles with low temperature or low thermal mass, usually small supercooled droplets or large supercooled drops with temperatures colder than -25 °C. These contact the ice particle and freeze almost instantly. The latent heat release does not exceed 0 °C, so microscopic air bubbles are trapped within the coating. This translucent, cloudy coating is known as rime ice.

By some sort of process within the cloud, the hailstone is alternately exposed to rime ice and clear ice This alternate exposure builds the characteristic, onionlike clear/rime layers seen in hailstone cutaways. During the 20th century, recy-

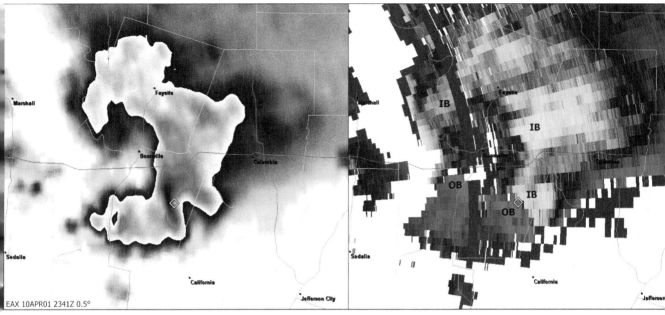

Figure 5-2. The most devastating hailstorm in U.S. history produced $1.5 billion in damage on 10 April 2001 in Missouri. The storm is seen here on radar reflectivity (left) and storm-relative velocity (right) with IB marking inbound and OB outbound velocities. The image was taken at 2341 UTC, showing the storm bearing down on Columbia MO. The storm shows strong HP signatures with evidence of a "bears cage" or radar notch outlining the updraft. Nearby at the diamond mark, the hail detection algorithm signaled a 4.2" maximum estimated hail size. The corresponding location on velocity shows a very energetic mesocyclone. At this 80 nm range, however, it is too far to reliably detect a TVS. A bookend vortex can be seen further north. Thanks in part to timely warnings only one fatality was reported.

cling of the hailstone in and out of the updraft was assumed to be the process responsible, but this idea has fallen out of favor. One school of thought holds that hailstones bubble at the top of the updraft, much like the suspension of lottery balls in an air mix machine. Shedding of droplets on hailstones is another source of hail embryos, as this instantly provides a source of particles 1 to 2 mm in size.

Like any liquid or solid object released into the air, a hailstone has a specific terminal velocity, or fall velocity, that is proportional to its size. Since this speed is frictionally based, it is relative to the surrounding air, not the earth, so if a hailstone has a terminal velocity of 50 $m \cdot s^{-1}$ and encounters an updraft with a 50 $m \cdot s^{-1}$ vertical velocity, the hailstone will appear to levitate at a constant height above the ground. Consequently strong amounts of instability intensify the updraft and directly support the growth of large hail. Sooner or later a hailstone grows sufficiently large that its terminal velocity exceeds the updraft velocity, and it falls to the ground. By this time it has usually spent at least 10 minutes growing in the cloud.

The growth of a hail embryo appears to work best along the edges of an updraft. This is thought to be because particles close to the core get carried into the anvil and away from the storm, while

Hail records

The largest hailstone fell 23 June 2003 in Aurora NE, measuring 7 inches in diameter and weighing 1.33 lbs.

However the heaviest hailstone fell in Coffeyville KS in 1970. It weighed 1.67 lbs and measured 5.7 inches in diameter. Until 2003 it was the largest known hailstone.

Green clouds

Greenish discolorations in the cloud have long been associated with the occurrence of hail. However a study published in Bull. Amer. Meteor. Soc. in the 1990s concluded that the discoloration was not due to hail but due to scattering of illumination through a very deep cloud.

Hail roar

The roar of hail has been documented anecdotally and even by chasers as far back as the early 1980s in issues of Stormtrack magazine. While little research has been done on this issue, the informal consensus is that the hail roar is indeed due to hail.

particles entirely outside the updraft fall to the ground. A hail embryo between these zones may escape these fates.

There is some speculation that large amounts of instability and shear may augment hail production by diversifying the possible range of trajectories in and around the updraft. Shear also produces dynamic pressure perturbations in the storm that may be involved in shaping these trajectories.

1.2. HAILSTONE FALL. The hailstone eventually makes its terminal fall. Once it reaches warmer layers in the lower troposphere it begins melting. The level at which melting occurs is not dictated by the freezing level, but by the wet-bulb zero (WBZ) height. This is the level at which the wet-bulb temperature is 0°C. It is normally at or just below the freezing level in an air mass with high relative humidity, but significantly below the freezing level in an air mass with low relative humidity. It should be pointed out that the WBZ height can vary considerably within the storm itself, adding to the difficulty of estimating melting characteristics.

If the warm layer near the ground is deep and warm enough, the hailstone may melt before it reaches the ground, turning into wet hail or rain, and even the rain itself might evaporate. But if the hailstone melting layer is shallow or nonexistent, then the hailstone will reach the ground at close to its maximum size. Regardless of the temperature profile, a large hailstone is significantly less prone to melting due to its faster fall velocity and its larger thermal mass, and its size is more a function of storm severity than air mass temperature and moisture.

Stroboscopic photography of small hailstones conducted by Matson and Huggins in 1980 confirmed very high spin rates, with most stones exhibiting about 40 revolutions per second on a horizontal axis. Much like a fastball thrown at a baseball game, spin gives the object unique aerodynamic characteristics. One such effect is a slower fall speed than would be expected of a similar object.

1.3. RELATION TO STORM STRUCTURE. Large hail tends to fall very close to the interface between the intense updraft and downdraft. In a supercell, where the forward-flank downdraft represents the primary downdraft, this also holds true. In the northern hemisphere, the largest hail is usually found north of the supercell mesocyclone. The rear-flank downdraft can also be a prodigious hailfall area, especially in HP modes where the radar echo evolves into more of a kidney-bean shape than a hook.

In a supercell, decades of observations have suggested that hail diminishes in size and intensity during tornadogenesis and then rises as the tornado dissipates. This has been tied to the life cycle of cyclic supercells and led to the concept of updraft collapse, suggesting that the weakening of an updraft correlates with the period of large hail fall. However this is now thought to be based on a rather simplistic model of an updraft as one large plume of rising air. Instead of an updraft collapse there are likely other processes at play at the time of tornado dissipation which, for other reasons, draw hail three-dimensionally to the surface. However the correlations are not well understood.

Hail shapes
The layering of the hailstone gives insight about how it formed. Conical, irregular hail is thought to result from riming on irregularly-shaped sources such as ice crystal aggregates and broken frozen drops.

2. Hail prediction

The science of predicting hail remains in its infancy. Much of the problem is rooted in the inability to directly observe hail particles within the cumulonimbus cloud, along with weak correlation of diagnostic parameters with actual hail sizes. Fortunately the widespread arrival of polarization diversity radars, which allows discrimination of storm regions by particle type, will likely allow significant advances to be made in this field during the 2010s.

Here we will examine the forecast methods for hail. Detection methods are almost entirely radar-based, and that will be treated separately in the radar chapter.

2.1. FAWBUSH & MILLER METHOD (1953). Historical methods are important since some of these techniques are still in use. As far back as the 1930s, it was recognized that a sounding with great positive area above the freezing level was conducive to hailstorms. The first rigorous hailstone forecast method by Fawbush and Miller examined the positive area on a sounding, bounded by three sides of a triangle: (1) the moist adiabat from the convective condensation level (CCL) to the -5 deg C level; (2) the pressure grid line at the -5 °C level; and (3) the environmental temperature trace from the CCL to the -5 °C level. The size of this triangle was proposed to

> **Predicting a hailstorm**
>
> The ordinary cold of the air in summer, at a height of 25,000 feet, or less, may be sufficient to produce a hailstorm wherever a warm current invades it from the lower air. But in order to produce a hailstorm of uncommon magnitude or intensity it seems necessary that masses of air widely differing in temperature and moisture, and possibly in electric state, should be suddenly and intimately mixed.
>
> FRANCIS A. R. RUSSELL, 1893
> *On Hail*

determine the hail size. Fawbush and Miller provided a nomogram with corresponding hail sizes.

2.2. FOSTER & BATES METHOD (1956). Another method, the Foster and Bates technique, came into use in the 1950s. It was similar to the Fawbush & Miller method, but considered buoyancy at the -10 °C level and other parameters. It used the relation $x = 3CHT'd/(4Td)$, where x is the hailstone diameter, C is the spherical hailstone drag coefficient, H is the depth of the positive area, T' is the difference in temperature between the parcel and the environment at -10 °C, d is the air density, and T is the environmental temperature at -10 °C. The method was adapted by Prosser and Foster in 1966 for early numerical model guidance, but later verification found that the technique had little skill.

2.3. WET BULB ZERO (WBZ) HEIGHT. The WBZ parameter examines melting influences on hail during the final fall to earth. With all other factors constant, a large distance should favor long melting time and greater melting of the hailstone. This principle was used in a revision of the original Fawbush-Miller method (Miller 1972). However the actual WBZ height the hailstone encounters varies depending on whether it is falling through an updraft or a downdraft region, and there is no certainty where that fall path will be. Furthermore, WBZ height yields no information about the degree of warmth of the warm layer, which further influences melting. Finally, the influence of WBZ is inversely proportional to hailstone size. Because of this, the usefulness of WBZ declines in unstable, sheared environments. The Edwards and Thompson study in 1998 found WBZ by itself to have little forecasting skill.

2.4. RENICK AND MAXWELL METHOD (1977). The Renick and Maxwell method provided more flexibility by examining the temperature and buoyancy at the level of maximum buoyancy, where the parcel achieves the greatest warmth relative to the environment at that level. The Moore and Pino 1990 method uses a forecast sounding algorithm that applies modification and examines the positive area above the LFC. It also takes into account surface heating, soil moisture, and low-level moisture, as well as water loading and entrainment effects on the updraft. This technique is largely focused on cloud dynamics and instability effects but does not take into account shear. SPC presentations have indicated that the technique shows little skill.

2.5. CONVECTIVE INSTABILITY (CAPE). The use of CAPE in hail forecasting is based on the correlation of instability with updraft speed. A basic rule of thumb suggests that storms will support hailstones with a terminal velocity of 50% of the updraft velocity. Though the CAPE contribution to vertical velocity can be accurately computed, the effects of shear add considerable uncertainty to instability-based estimates. These greatly influence trajectories and residence times of hail embryos. Unfortunately verification of CAPE and CAPE derivatives by Edwards and Thompson 1998 has shown that they are poor predictors when used by themselves.

2.6. SIGNIFICANT HAIL PARAMETER (SHIP). This diagnostic parameter was developed around 2000 by SPC and can be condensed to $X = B_u w_u \gamma_m (-T_m) V / 4.4 \times 10^7$, where X is the hail parameter, B_u is most unstable parcel CAPE (J·kg^{-1}), w_u is most unstable parcel mixing ratio (g·kg^{-1}), γ_m is the 700-500 mb lapse rate (°C·km^{-1}), T_m is 500 mb temperature (°C), and V is the 0-6 km shear in m·s^{-1}. Since this is a multiplicative index, an increase in any one of these parameters enhances the index, except for the inverse relationship of 500 mb temperature. This index is not meant to forecast hail size but specifically to signal if large hail (0.75 in or greater) is likely. This occurs when the SHIP parameter gives a result of 1 or more.

2.7. ENERGY SHEAR INDEX. One study by Brimelow and Reuter 2002 indicated success with multiplying CAPE by 1.5-6 km MSL mean vertical shear magnitude to obtain the energy shear index (ESI, m^2·s^{-3}). This was proposed as a way of parameterizing updraft duration. An ESI value approaching 5 is considered favorable for large hail, with values above 5 not having further significance. The method, used in conjunction with model output, was adopted by SPC in 2001 and has been quite successful in reliably identifying large hail environments.

Supercooled water

In the Vienna Sitzungsbericht, February 1900, just received at the Weather Bureau, Dr. P. Czermak details some experiments on the cooling of water below its freezing point and the forms that can result when it suddenly freezes. He concludes that not only can water, cooled below freezing, form a cloudy kernel of mixed ice and rain, but also water that has not been thus subcooled, and that, too, in a very deceptive manner; in many cases the opaque grains or nuclei of hailstones must have been formed of water that was not cooled below zero, centigrade, before it froze. Similar remarks apply to crystals of snow which may, therefore, have been formed at the freezing temperature without requiring any previous cooling to lower temperatures. Undoubtedly some hailstones are formed without these opaque nuclei.

CLEVELAND ABBE, 1901
editor, *Monthly Weather Review*

3. Hail detection

Hail detection methods are based almost entirely on radar signatures. Therefore this topic is covered in depth in the radar chapter. At some augmented weather stations, observers may report and transmit hail sizes, but due to the automation of the federal observation network these reports are far more rare than they were in the 1980s. ¤

Table 5-1. Hail size category table.

Inches		mm	Description
1/8"	0.10"	0.3 cm	Bird shot[3]
1/4"	0.25"	0.5 cm	Pea
1/2"	0.50"	1.5 cm	Grape, mothball, marble[1]
3/4"	0.75"	1.9 cm	Dime, penny, marble[1]
7/8"	0.88"	2.2 cm	Nickel
1"	1.00"	2.5 cm	Quarter, walnut, loonie[2], toonie[2]
1 1/4"	1.25"	3.2 cm	Half-dollar
1 1/2"	1.50"	3.5 cm	Ping-pong ball
1 3/4"	1.75"	4.4 cm	Golf-ball
2"	2.00"	5.1 cm	Hen's egg[3]
2 1/2"	2.50"	6.5 cm	Tennis ball
2 3/4"	2.75"	7.0 cm	Baseball
3"	3.00"	7.6 cm	Tea cup[3]
4"	4.00"	10.2 cm	Grapefruit, melon
4 1/2"	4.50"	11.0 cm	Softball

[1] Marble size covers a range of 0.5" to 0.9" and its use is discouraged. It is provided here solely for historical use.
[2] Canadian loonie ($1) and toonie ($2) coins have diameters of 2.65 cm (1 in) and 2.8 cm (1 1/8 in), respectively.
[3] Not a widely recognized measure.

6

LIGHTNING

Many hypotheses have been propounded to explain the origin of atmospheric electricity. Some have ascribed it to the friction of the air against the ground, some to the vegetation of plants, or to the evaporation of water. Some again have compared the earth to a vast voltaic pile, and others to a thermo-electrical apparatus. Many of these causes may in fact concur in producing the phenomena.

— Adolphe Ganot, 1868
Elements de Physique

LIGHTNING

Lightning holds a special place in the history of human culture, its worldwide distribution shaping the legends of many cultures. Its awesome power was ascribed to the god Zeus in ancient Greece, the goddess Fulgora in ancient Rome, and the Thunderbird in the Americas, a creature which was recognized universally across most Native American tribes. The Bible holds that God's power and presence was sometimes accompanied by displays of lightning. And even science itself suggests that lightning was literally the spark of life on Earth, turning water, hydrocarbons, and ammonia into complex amino acids.

Though ancient cultures knew little about lightning, it was well known that amber when rubbed with dry animal fur could attract hairs and cobwebs. It was soon found that this electrostatic effect caused lodestones to respond. Its link with thunderstorms was first explored in the 1752 kite experiments by Thomas-François Dalibard and, apocryphally, Benjamin Franklin. Though the science of electromagnetism made rapid strides after 1820, a comprehensive understanding of lightning, would not emerge until the 20th century when relationships between updrafts, downdrafts, water droplets, and ice crystals were more fully understood. Only then did technology such as cameras, aircraft, field mills exist to allow verification of the theories.

1. Lightning types

Lightning itself does not have any known effects on weather. It is primarily an indicator of storm strength and a factor in the safety of people and property. Each year, lightning causes an average of 83 deaths and up to $6 billion in damage in the U.S., according to some estimates. Early in the 20th century it was speculated that tornadoes were powered by lightning, but these theories are widely discredited.

1.1. CLOUD-TO-GROUND LIGHTNING (CG). Most commonly this flash occurs between the negatively charged mid-levels and the ground. Negative strikes are usually the result of the convective precipitation core interacting with the ground. A less common type of flash occurs between the positively-charged upper levels of the storm and the ground, often referred to as a "positive CG", described in more detail shortly.

1899: What was known

The quantity of electricity in the cloud is continually increased by the inflow of moist air at its lower margin. It is believed that the electric potential is increased by the aggregation of many extremely small cloud particles into a smaller number of larger droplets, and finally into rain-drops; for the initial charge resides on the surface of each minutest particle, and with the successive aggregation of particles, the quantity of electricity increases faster than the surface area of the droplet.

Thus with growth of the cloud, there is both increased potential and increased quantity of electricity. It is probable that this process goes on in all cases of cloud formation; but that a potential high enough to cause lightning flashes is produced only when the cloud growth is rapidly and continually augmented by inflow of moist air at the base.

WILLIAM M. DAVIS, 1899
Harvard scientist

Figure 6-1. Cloud-to-ground lightning beneath a severe multicell storm, 28 May 1997, near Earth, TX. *(Tim Vasquez)*

Harnessing lightning

When people learn about the thousands of volts of electrostatic charge that make up the fair weather field, it's often suggested that this energy source ought to be tapped to produce electricity. Unfortunately air is an extremely good resistor. An atmospheric probe would allow electrons to flow only from the atoms that collide with the probe, producing a negligible amount of current. It would not be able to tap the current throughout the free atmosphere.

Storms generate vastly larger amounts of voltage and current, but this is concentrated along very specific pathways that are difficult to predict. A lightning strike on a radio tower might be harnessed, but there are technological problems in storing this energy and feeding it to the grid. There is also the question of yield. A lightning strike only produces about 200 kWh of electricity, roughly the amount of power a large coal-fired power plant produces in eight seconds.

1.2. CLOUD-TO-AIR LIGHTNING (CA). Often seen with the strongest storms, cloud-to-air lightning typically reaches from the positively charged upper regions of the storm into surrounding clear air. Historically this was sometimes called *rocket lightning* due to its tendency to discharge upward above the cloud. There is evidence that a few rare strikes may even extend into the stratosphere; this type of lightning may be related to blue jets (q.v.).

1.3. CLOUD-TO-CLOUD LIGHTNING (CC). Lightning may reach between two different convective clouds, sometimes stretching as far as 100 miles and most frequently through the anvil. The anvil of a storm is often heavily electrified in severe storms, producing discharges dozens of miles long called *anvil crawlers*, and in the MCS such anvil discharges can spark between numerous convective towers.

1.4. INTRACLOUD LIGHTNING (IC). Lightning that occurs within the thundercloud itself is considered "in-cloud" or "intracloud". Although it is very frequent, it is rarely seen, but at nighttime it is quite effective at illuminating the cloud from the inside.

1.5. NEGATIVE CG DISCHARGE. Well over 90% of cloud-to-ground lightning strikes take on negative polarity; that is, they transfer charge from the negatively-charged cloud to the ground. The textbook description of a lightning flash details this type of discharge. It contrasts from the positive CG discharge, described below.

1.6. POSITIVE CG DISCHARGE. Lightning flashes may occur which transfer charge from the ground upward to positively-charged portions of the thunderstorm. Positive CG lightning is typically produced by severe storms (particularly the trailing portion of an MCS), the dissipation stage of any storm, and cold-season storms. Positive strikes bear association with a special type of lightning known as a sprite, which will be discussed shortly.

Surprisingly, positive flashes are more common in the cool season, with December being the most common month for +CG strikes in the United States. This correlation is associated with ice crystals and upper tropospheric regions being closer to the ground, along with an increase in stratiform inverted dipole characteristics. Increased winds aloft, which are more common in the winter months, are more effective at shifting positive charge horizontally where they lie above the ground without being shielded by the negatively-charged precipitation core.

2. Lightning formation

The science of storm electricity is based largely on electromagnetic theory and is well outside the scope of this book. In fact as a scientific discipline it lies partly outside the sphere of meteorology and some of the research is not even comprehensible to highly educated meteorologists. Our intent in this chapter is to summarize the key findings in lightning research as it relates to operational forecasting. While the lightning flash itself is well understood thanks to the advanced state of physics, the underlying meteorological processes and their exact relation to clouds and weather remain areas of intense study.

2.1. ELECTROSTATIC CHARGE. Lightning is nothing more than an electrical arc between two intense charge centers. To understand charge, consider that matter such as water, air, and dust is made

Jets, sprites, and elves

Between 1989 and 1995, researchers documented and identified a special class of lightning-like phenomena occurring in the upper atmosphere above thunderstorms. These phenomena had been suspected for almost a century.

Blue jets are the most visible form of this phenomena and can be seen at night by the unaided eye. Blue jets appear as beams of bluish light lasting about half a second which radiate from the top of the storm up to about 40 km. They occur independently of lightning flashes within the cloud, and there is evidence that they discharge high positive cores which are not sufficiently discharged by +CG strikes.

Sprites sometimes occur in the ionosphere above strong thunderstorms at an altitude of about 50 to 100 km. They are usually too brief and faint to be seen with the naked eye, but brighter discharges may be visible. It is thought that they are caused by very powerful cloud-to-ground lightning strikes, which in turn briefly disrupt the electrical field above the storm.

Elves are an extremely brief glow that occur in the ionosphere (about 70 to 100 km MSL), lasting about one millisecond. The glow can measure as wide as 500 km. They are thought to result from electromagnetic pulse of flashes within the thunderstorm which cause ionospheric gases to glow.

up of atoms and molecules, all of which are made up of a nucleus of protons and neutrons surrounded by an outer shell of electrons. Normally the number of electrons and protons are equal, but if some process disrupts this balance and removes or adds an electron, the atom or molecule is becomes an *ion*. If the ion has an electron deficit, it is considered to be *positively charged* and if it has a surplus it is *negatively charged*.

Charge centers can exist in the atmosphere for long periods of time because air is an extremely good insulator, so the phenomena is electrostatic in nature. If the charge centers are extremely intense and close enough to each other, the voltage gradient can exceed the air's ability to provide electrical resistance. This breakdown occurs at roughly 1 million volts per meter. It produces an electric arc which ionizes the air along a narrow pathway, turning it into a plasma, and allows the surplus electrons to flow from the negative to the positive charge center. The result is a lightning flash.

2.2. FAIR WEATHER FIELD. The earth's surface has an electron surplus which comes from the hundreds of thunderstorms in progress all over the earth, and this electron surplus gives it a negative charge. The upper atmosphere takes on positive charge, in part from the effects of thunderstorms, and this distributes upward to the ionosphere in the form of leaks and other phenomena such as sprites and elves. As conduction of the air is much higher at this extreme level, the positive charge is distributed rapidly in the ionosphere across the entire globe. Since opposite charges attract, free positive ions in the atmosphere are attracted downward and accumulate in the lower troposphere, particularly in the lowest kilometer. These free positive ions come into play later.

2.3. THUNDERSTORM DEVELOPMENT. As a thunderstorm grows, the inflow and updraft collects the layer of positive ions near the earth's surface and moves them upward through the atmosphere. The top of the cloud accumulates significant positive charge. This area is commonly referred to as the upper positive, or **P**, region. It is reinforced as ice crystal production begins. The upper positive region attracts negatively charged ions from the upper troposphere, forming a skin known as a "negative screening layer".

Early lightning experiment

Several gentlemen interested in scientific pursuits engaged upon trials [of Benjamin Franklin's experiment], among them two persons of note, M. Dalibard and M. de Lor. A wealthy man of science, Dalibard [lived] during a part of the year in a handsome country house situated at Marly-la-Ville, about eighteen miles from Paris ... A wooden scaffolding was built up to hold in its midst an iron rod, eighty feet long and slightly over an inch in diameter ... the rod entered, five feet from the ground, into a thinner one, running horizontally toward an electric apparatus, fastened to a table in a kind of sentry-box, erected on purpose for observations ... Dalibard was called by business to Paris, and left the whole in charge of one of his servants, an old soldier, formerly in the French dragoons, Coiffer by name ... There he sat May 10, 1752, when a violent thunderstorm drifted over the plain of Marly. He touched the electrical apparatus with a key, silk-bound at the handle, and to his extreme surprise sees a flame bursting forth. He touches another time and there is a second flame bursting forth, stronger than before. Then the old dragoon rushes from his sentry-box — most famous private dragoon that ever lived, born to the high honour of being the first man that ever drew lightning from heaven.

RICHARD ANDERSON, 1880
Lightning Conductors

LIGHTNING 115

Precipitation processes which we will describe shortly cause the middle levels of the cumulonimbus cloud to accumulate negative (-) charge. This level is referred to as the main negative, or **N**, region. This area is concentrated around the -10 to -20 °C layer (usually at about 15,000 to 25,000 ft MSL in a typical late spring storm). It is caused largely by the effects of graupel - ice crystal collisions.

The prevailing configuration of opposing charge between the P and N layers is often referred to as the "thunderstorm dipole". When the positive charge exists above the negative charge, this is referred to as a *normal dipole*, though in certain situations the charge regions can be upside down, producing an *inverted dipole*.

There is an additional charge area that bears mention. The area near the cloud base accumulates modest positive (+) charge. It is referred to as the lower positive charge, or **p** region. This positive charge comes from a corona effect on the earth's surface caused by the main negative region lying overhead. The positive area is also created from free positive ions in the lower troposphere that are attracted to the negative region. This lower positive charge region is

Absurd but entertaining

"I was staying at Stockton Heath, England, in July 1919, about 400 yards from the Manchester Ship Canal. The evening was somewhat oppressive, and the air had become strangely still. Gazing down the road I saw a small black thundercloud gathering along the length of the canal, and about 30 or 40 feet above it. It was approximately 400 yards long and perhaps 6 feet thick.

"As I gazed at this strange formation, a dazzling lightning flash raced through the entire cloud, i.e. parallel to the water, and a bang like the discharge of field artillery followed immediately. About 40 seconds later, another flash and report occurred; then the cloud dispersed in about four minutes.

"I might add, that at least in those days an air current of varying intensity moved up that canal almost incessantly, i.e. inland toward Manchester, at the Stockton Heath section, one felt it on the neighboring bridge."

C. S. BAILEY, M.A.
Weather, 4:267, Royal Met. Soc.

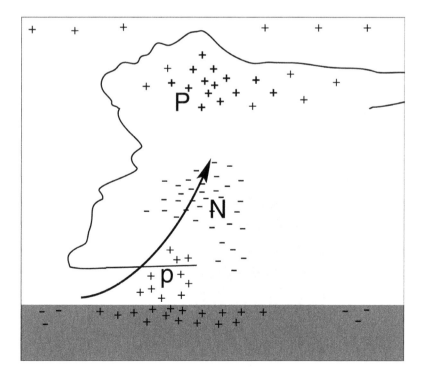

Figure 6-2. Thunderstorm charge centers in a cumulonimbus cloud. The negatively charged core in the mixed region is marked N, the positive ice crystal region P, and the positive subcloud region is p. *(Tim Vasquez)*

The lightning flash

Lightning begins when the field in a particular area of the storm strengthens locally, overcoming the electrical resistance of the air and causing dielectric breakdown.

When the air reaches breakdown strength, a stepped leader moves quickly downward to the earth. This leader ionizes the air and produces a conductive path.

A very bright, powerful return stroke moves upward and transfers charge. The stroke is very hot, on the order of 30,000 deg C, and this produces a shock wave heard as thunder.

This cycle may occur 3 or 4 times, with a new step leader known as a "dart leader" that prepares the channel for another return stroke.

Note that the term flash or discharge refers to the entire event, while the term strike or stroke applies to one individual bolt.

Once lightning occurs, it is generally assumed that the flash neutralizes the opposing charge centers. Some studies however have found that lightning can increase charge within the cloud.

---

collected as storm inflow and may be lifted a short distance into the storm updraft.

Therefore the thunderstorm dipole, when occurring together with the lower positive charge region p is referred to as a *tripole configuration*. The axis of the dipole or tripole can tilt. This usually occurs in a sheared wind environment when the upper positive charge center is advected downwind by upper level winds relative to the main negative center. A tilted dipole or tripole essentially removes the blocking effect of the middle negative charge center, permitting positively-charged lightning strikes between the upper positive region and the earth's surface.

2.4. PRECIPITATION PROCESSES. The many different collision types and interactions are a key to understanding the basic principles of how thunderstorms become charged.

The most basic precipitation process is collision and coalescence, where liquid droplets collide with one another and grow. Initially the droplets are very small and relatively very far apart from one another, and collisions are very rare. Furthermore if the droplets are all the same size, they all have the same terminal velocity and thus fall at the exact same rate as one another. However if enough collisions occur to produce numerous droplets of different sizes, this produces a range of different fall speeds. When this happens, much like a freeway where some cars are driving 100 mph and others at 20 mph, many more collisions will occur, causing a chain reaction. This cascade may also be caused by *collector drops* falling into this layer from different parts of the cloud or from different clouds altogether.

In charged clouds the *electrostatic effect* comes into play. Each droplet develops a negative and positive charge center on its surface in response to the dipole within the cumulonimbus cloud. Since the droplet now exhibits a charge, many of the droplets begin attracting one another, greatly accelerating the collision and growth process.

Then we must consider supercooled liquid. It is possible for both liquid and ice to co-exist in a stable configuration, without any change in state, if temperatures are between 0 deg C and about -42 deg C. In a typical storm on the Great Plains, this mixed layer of supercooled water and ice spans a considerable depth between roughly 3 and 12 km MSL. The *Bergeron-Findeisen process* allows ice growth in this mixed layer at the expense of supercooled water. This is because vapor pressure over water is greater than that over ice, and as a result water vapor evaporates off the supercooled drops and condenses on the ice.

Eventually there is interaction between the supercooled droplets and the ice crystals. If an ice crystal collides with a supercooled droplet, the droplet flash-freezes, causing *accretion* of rime ice on the crystal and forming a new type of particle known as graupel.

2.5. PRECIPITATION CHARGING. Graupel is one of the key elements in thunderstorm electricity. Graupel is the name for a small water droplet larger than a cloud droplet and smaller than a rain drop, which has frozen. This frozen particle is not merely from the drop moving into subfreezing layers, since a liquid drop easily remains in the supercooled state, but rather from interaction with an ice crystal or a snowflake. The frozen drop is called graupel.

When this freezing process occurs, the basic electromagnetic properties of the particle causes some electrons to move into the liquid core. As a result, the frozen shell acquires a positive charge, while the liquid core takes on a negative charge. Once the core itself freezes, the density change causes it to expand volumetrically, causing the particle to crack or shatter. This casts off small fragments of ice which are electron-deficient, carrying positive charge, and due to their very weak fall speed they are swept up by the updraft into the upper portions of the storm, building positive charge there. Meanwhile the negatively-charged graupel core is heavy and descends to lower elevations of the storm. The creation of this vertical configuration is an extremely important process and tends to form the normal dipole arrangement within the storm.

As the negatively-charged graupel descends closer to the ground, it induces positive charge at the ground. Electrical breakdown as this negative core gets closer to the ground results in the production of negative cloud-to-ground flashes between cloud and ground.

Once special note worth mention — in the lower parts of the storm where temperatures are warmer than -15 deg C (below about 20,000 MSL in a typical Great Plains atmosphere), a graupel-ice crystal collision actually moves electrons to the ice crystal and gives

Lightning distance
The proximity of the strike can be estimated by the long-time rule of thumb, based on the speed of sound, in which the delay in the thunder following the lightning flash indicates a distance of 1 mile for every 5 seconds (1 km for every 3 seconds).

Lightning dangers
About 75 deaths occur each year in the United States due to lightning. There are a number of safety rules for lightning. Seek shelter in a car or building if a thunderstorm is around. If a tingling sensation is felt, a sizzling sound is heard, or hair stands on end, a lightning strike is imminent and one must immediately crouch down.

Lightning lexicon

Provided here is a list of key phrases used in lightning detection technologies.

Flash count. The total number of strikes within a specific area and time period.

Flash rate. The total number of strikes within a specific area and time period, divided by the time length for that period. Flash rate is largely a function of rate of charge separation, density of the cloud, distance between the cloud and ground, and degree of shielding of charge centers.

Flash density. The flash density of a storm is the total number of strikes within a specific area and time, divided by the geographic area.

Negative strike dominance (NSD). A storm which shows a predominance of negative strikes, which is normal for most storms.

Positive strike dominance (PSD). A storm which has a predominance of positive strikes, usually associated with severe modes.

---

the graupel positive charge. The negatively-charged ice crystal rises while the positively-charged graupel sinks. This promotes an inverted dipole, which in fact reinforces the **p** region of the storm.

Due to the thermo-electric properties of water molecules, even particles with *different temperatures* can carry away different charges when they collide. Warm particles tend to gain negative charge while cold particles gain positive charge. What this means is that warm hailstones entering a zone of cold ice crystals will become negatively charged. Particles with *different size* can acquire different charges, too: for example a large particle has a tendency to acquire negative charge and the small one positive. All of these factors reinforce the positive charge of small ice crystals.

It can be seen that graupel and ice are extremely important in producing lightning. In fact, they are so important that warm-cloud lightning, that is, caused by a cumulonimbus cloud with no portion below freezing, has been a long-standing issue of debate in the meteorological community with some researchers maintaining that warm-cloud lightning is impossible. Observational evidence is still inconclusive and though some degree of cloud electrification is supported in a warm cloud, no process has yet been found that appears to be strong enough to cause electrical breakdown.

3. Dipole characteristics

The dipole created in the cumulonimbus cloud by ascending, positive ice crystals and descending, negative graupel is an important property of the thunderstorm. It is described in more detail here:

3.1. DIPOLE TILT. The vertical axis of the thunderstorm dipole may tilt due to wind shear between middle and upper regions. For example, as the positively-charged ice crystals near the top of the storm begin moving with stronger upper-level winds, it tends to spread or carry the upper positive region downstream from the storm while the negative core stays behind. This produces a tilt in the dipole arrangement. It also brings the positive charge center over the ground without the shielding of the main negative area in between. This may produce positive cloud-to-ground strikes.

3.2. MCS LIGHTNING. After a squall line passes, it is common for a trailing stratiform region to follow the line, bringing overcast skies, light rain, and a few powerful positive cloud-to-ground strikes. How does this occur?

Abundant growth of supercooled droplets does not occur in a stratiform region. Graupel and hail are mostly absent and the stratiform layer is mostly filled with ice crystals. When ice particles collide, they aggregate into clusters. Collisions between ice crystals and ice crystal aggregates has the effect of transferring positive charge to the aggregate and negative charge to the ice crystal. Relatively speaking, the heavier aggregate falls and the lighter ice crystal rises. This produces an inverted dipole in the trailing stratiform region. The result is that a positive layer of ice crystal aggregates develops about 3 to 8 km above the ground. As a result, positive cloud to ground strikes tend to occur.

4. Lightning detection systems

Since lightning strikes emit a powerful electromagnetic pulse, there are specialized sensors available which can detect the activity.

4.1. ELECTROMAGNETIC DETECTION. Professional, real-time lightning detection equipment uses wideband magnetic direction finders, each with a range of about 400 km. The detector reports azimuth, amplitude, and polarity of the strike. This does not yield distance, though, so a network of sensors is required, which allows location to be determined by analyzing the strike azimuth at multiple sensors or deducing location from time of arrival at each sensor.

In the United States, the National Lightning Detection Network (NLDN) is the primary lightning detection network for forecast use. It was established in 1987, combining the BLM network in the west U.S., the NSSL network in the central U.S., and the SUNY Albany network in the east U.S. Currently it is operated by Global Atmospherics, consisting of about 100 stations located nationwide. Unfortunately, due to the network being privately operated, the general public and hobbyists receive somewhat degraded forms of output.

Electromagnetic detection technology has also been miniaturized into handheld "personal lightning detectors", but since these

---

Lightning superbolts

Unusually intense lightning strokes have been observed by optical sensors on the Vela [nuclear test monitoring] satellites. These lightning flashes are over 100 times more intense than typical lightning. The lightning superbolts are characterized by optical power in the range of 100 billion to 10 trillion watts, have a duration on the order of 1 ms, and have a total radiant energy of 1 billion J. In conjunction with sferics data the Vela trigger rates indicate that about two lightning flashes in 1000 exceed an optical power of 10 11 W and five flashes in 10 million exceed an optical power of 3 trillion watts.

B.N. TURMAN, 1977
Detection of Lightning Superbolts

Thunder

The rumbling of thunder is caused by different parts of the lightning channel reaching the observer at further slant ranges, and echo off buildings, hills, and trees can add to the effect. Thunder may be heard up to 25 km away, but rarely any further since sound has a strong tendency to refract upward, limiting its audible range. This has created the name "heat lightning" in which distant storms are seen but no thunder is heard.

devices rely on careful positioning and don't measure the electrostatic field, debate exists whether these devices are suitable for assuring lightning safety at a given spot.

4.2. OPTICAL DETECTION. Starting in 1995, NASA sent optical lightning detectors into space aboard polar orbiting satellites. The problem with these is that there is no real-time display of data, and since the satellites are not geostationary they move swiftly around the earth, observing a given spot only once per day for a few minutes. Though the detectors are extremely valuable for research purposes, the systems are not practical for operational forecasting use.

However the Geostationary Lightning Mapper (GLM) will appear in the GOES-R series of satellites, due for launch in 2015. It uses a near-infrared optical transient detector 777.4 nM to continuously detect lightning activity, achieving a resolution of 8 km. This will not provide the precision of NLDN but will serve as an excellent optical-based alternative source of this data, and will provide excellent coverage for storms offshore, such as hurricanes. Products will be distributed through the NOAA product suite.

4.3. RADAR DETECTION. Studies have underscored the value of predicting lightning based on the growth of moderate reflectivity values into specific temperature layers, as determined by a nearby sounding. In essence, this determines whether a core of mixed supercooled water and ice might exist. Michimoto in 1990 found that when reflectivities increase to 30 dBZ at the -20 deg C level in a developing storm, lightning follows about five minutes later. A similar technique suggests using 40 dBZ values at the -10 deg C layer (Buechler and Goodman, 1990), which gives about 7 to 8 minutes of leadtime.

On rare occasions a radar unit can directly detect electromagnetic radiation from a lightning strike, either detecting the flash or the ionization channel. However no established techniques exist that allow forecasters to rapidly discriminate between these anomalies and other types of interference.

4.4. FIELD MILL. Finally, another type of technology is the field mill, which simply measures the strength of the electrostatic field at the sensor's location and examines changes over time. It does not specifically detect lightning. Very strong electrical fields tend to be associated with charge centers overhead in thunderstorm clouds, giving some indication whether lightning might be possible at the field mill's location. The systems are often used at airports and factories for safety assurance, and are usually good at warning for an imminent lightning threat.

5. Lightning forecasting

As is indicated by the radar technique above, the detection of 40 dBZ echoes penetrating to -10 deg C or 30 dBZ echoes penetrating to -20 dBZ is a good signal that a convective cloud is likely to produce lightning. These strikes are initially in-cloud and may not be detected well at first by electromagnetic lightning networks.

Storms which strengthen usually exhibit an increase in flash rate and flash density. These quantities are readily estimated from electromagnetic devies. Polarity is an important characteristic, particularly in areas where radar data is unavailable. Storms are usually negative strike dominant. However an increase in positive flash rates may show that strong upper level winds have blown the positively-charged anvil downwind. Very strong anvil-top divergence can also cause a backsheared anvil, that is to say it spreads out in a mushroom shape, and this can also spread the upper positive area over areas not shielded by the negative charge region and cause +CG strikes there. Such strikes from backsheared anvils are usually confined to within about 5 or 10 miles of the storm.

Some preliminary work is beginning into model forecasting of lightning. Though individual storm paths cannot be predicted, much less the location of individual strikes, favorable conditions in the storm *environment* can be predicted. Experimental lightning prediction forecasts have been developed for the WRF (McCaul et al 2007) which measures forecast ice flux in layers near -15 deg C.
⌑

Lightning climatology

The United States may reign supreme when it comes to tornado and hail activity, but the title for lightning goes to the Democratic Republic of the Congo. A NASA spaceborne sensor known as LIS/OTD has shown the world's lightning hot spot to be in the central and eastern Congo: home of Conrad's Heart of Darkness and spanning much of the world's mountain gorilla population. In the United States, Florida takes the title as America's lightning capital. The incidence gradually decreases as one moves northwestward from Florida through Louisiana and into the eastern Great Plains. Interestingly, older maps show the nation's high lightning concentration to be in northeast New Mexico, but this is not supported by current data. Since human observational data makes up the backbone of early records, observational bias at Clayton are thought to be the cause.

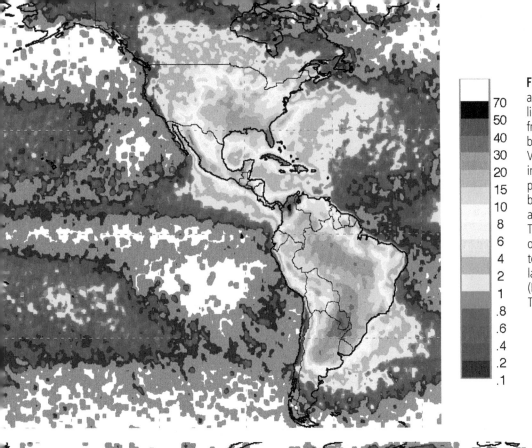
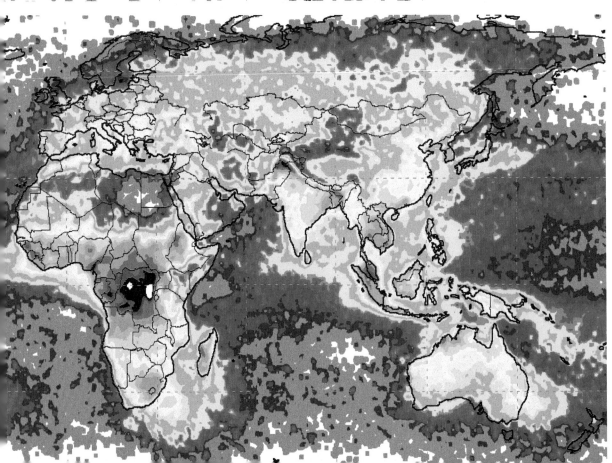

Figure 6-3. Worldwide annual distribution of lightning, as obtained from the NASA spaceborne LIS/OTD sensor. Values are expressed in mean annual flashes per square kilometer between April 1995 and December 2005. The lightning capital of the world appears to be the eastern highlands of central Congo. (NASA LIS/OTD Science Team)

7

STABILITY & SHEAR

The forecasting of all types of thunderstorms
depends primarily upon the use of upper air data
obtained from aerographic soundings. Without this
information it is impossible to know the temperature
and moisture distribution in the upper atmosphere
and thus it is quite impossible to know anything
about stability conditions.

— George F. Taylor
Aeronautical Meteorology, 1938

A thunderstorm simply cannot exist unless large quantities of moist cloud material are moving upwards rapidly. This motion not only allows cloud material and precipitation to occur but also promotes electrostatic charge separation that produces lightning and creates unusual circulations that lead to hail and tornadoes.

For rapid updraft motion to occur, there must be a sufficiently moist low-level environment and strong upward acceleration of air parcels. Upward acceleration, in turn, is caused by instability. To measure instability, meteorologists must examine the distribution of temperature in the atmosphere using a chart called a sounding. Research in the past few decades has also underscored the importance of shear and helicity. Though these wind properties can be plotted on a sounding, they're best analyzed on a different kind of chart called the hodograph. This tool will be discussed later in the chapter.

1. Types of instability

Instability describes any condition where any displacement will result in the conversion of potential energy to kinetic energy, causing further acceleration in the direction of displacement. Likewise, stability resists displacement and causes the parcel to return to its starting position.

1.1. GRAVITATIONAL INSTABILITY. In an atmosphere with gravitational instability, an air parcel that is displaced in any vertical direction will be accelerated further in that direction. It occurs when the lapse rate is high; that is, when the environment cools rapidly with increasing height. This type of instability is the basis for convection.

1.2. INERTIAL INSTABILITY. If inertial instability is present, a displacement in any horizontal direction will result in further acceleration in that direction. It occurs when the absolute vorticity is negative.

The use of indices

In my experience, many forecasters, implicitly or not, are seeking a "magic bullet" when they offer up yet another combined variable or index for consideration. If forecasting were so simple as to be capable of being done effectively using some single variable or combination of variables, then the need for human forecasters effectively vanishes.

CHARLES DOSWELL, 2006

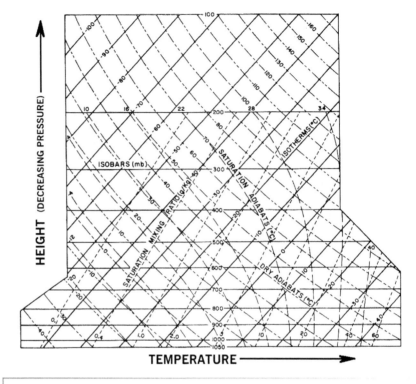

Figure 7-1. **SKEW-T log p diagram schematic.**, from Robert C. Miller's 1972 SKEW-T paper. This shows that the skew-T is simply a way for plotting an observation of weather data given its temperature and height (i.e. pressure). By taking many observations in the atmosphere above a given station, as through a radiosonde observation, we can connect these points to form a thermodynamic sounding. *(Miller)*

1.3. SYMMETRICAL INSTABILITY. Symmetric instability describes an atmosphere that is stable to both horizontal and vertical displacements; in other words, a parcel will return to its original level. But if it is displaced in a vertically slanted path, it accelerates further in that direction. It is occasionally responsible for weak shower and thunderstorm activity near frontal zones. The process will be described further in this chapter.

1.4. POTENTIAL INSTABILITY. Potential instability is the name given to potential gravitational instability, a thermodynamic condition that exists when the lower portion of a layer contains latent heat not yet released, while the upper portion does not. With a source of lift, the lower-level latent heat is released and results in a strengthening of lapse rates. An atmosphere with potential instability exists when the equivalent potential temperature value decreases with height in the layer.

1.5. BAROCLINIC INSTABILITY. Baroclinic instability is a type of horizontal and vertical instability that is caused by the contrast of air masses with different temperature. This is the basis for frontal systems. It will not be covered in this section.

2. Sounding basics

In its simplest terms, a sounding is nothing more than a x-y graph of temperature versus height above a given station, with an additional plot for humidity also added. Temperature and humidity are both a function of the x-axis, while height or pressure is a function of the y-axis. The ground is always at the bottom of the diagram and the upper atmosphere is at the top. With this in mind, it's possible to plot the tropospheric conditions at a given station at a specific time.

There are actually several types of graphs that are used for the sounding, all of them similar. The basic x-y representation of temperature versus height is used in the *emagram* and the *Stüve diagram*, but they fell out of use by the 1950s. In the United States and most severe weather research, the *skew-T log-p diagram* is widely used, which skews the temperature lines upward and to the right rather than vertically. This makes plotted temperature profiles easier to use by making them stand more erect. Forecasters in Canada and the United Kingdom use the *tephigram*, which is in fact the basis for the skew-T log-p and is very similar but uses curved pressure lines to more accurately relate geometric area to energy.

3. The skew-T log-p diagram

The skew-T log-p diagram was first proposed by Norwegian meteorologist Nicolai Herlofson in 1947 as a refinement to the emagram. This type of chart was quickly adopted by the U.S. Air Force and later the U.S. Weather Bureau. By the 1960s it became the favored diagram for atmospheric studies in the U.S. weather community. There are five sets of grid lines that form the skew-T log-p diagram, some of which are also applicable to the tephigram:

3.1. PRESSURE GRID LINES. Height or altitude is not a scientific coordinate system when it comes to thermodynamic work, so pres-

128 STABILITY & SHEAR

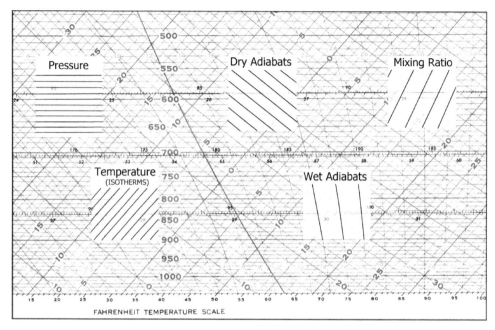

Figure 7-2. SKEW-T diagram and reference key to each of the five lines. Learning to identify the five line sets is perhaps the most important skill for novices. The wet adiabat lines can always be distinguished from other lines because they have curvature. The mixing ratio lines are sometimes another source of confusion, but they are usually drawn as a dashed line, are confined to the lower part of the diagram, are straight, and stand more vertical than the temperature lines. *(Tim Vasquez)*

sure is used as the key vertical coordinate. This is easy to understand since pressure falls rapidly with height. Each line represents constant pressure, *p*, in millibars (mb), also known as hectopascals (hPa). In a typical atmosphere where the sea-level pressure is 1013.2 mb, sea level would be found just below the 1010 mb line. Most soundings top out at 100 mb, which is roughly equivalent to 53,000 ft MSL, the typical upper limit of the troposphere.

A parcel does not follow the pressure lines, but can be measured against them.

3.2. TEMPERATURE GRID LINES. If temperature above a given station was plotted on standard graph paper, the trace would lean to the left, since temperature usually falls with height. This lean makes it harder to see important details. The key feature of the skew-T log-p and tephigram is that they "skew" the temperature grid lines so that they slope to the right rather than stand upright. When plotted against these skewed temperature parallels, the sounding trace stands upright. It takes a little practice to read skewed temperature lines, but it's easy when it's kept in mind that they always slope up and to the right. They are usually colored brown and are straight. Each line represents a constant value of temperature, T, in degrees Celsius.

A parcel does not follow the temperature lines, but can be measured against them.

3.3. DRY ADIABAT GRID LINES. Dry adiabats are slightly curved grid lines which slope up and to the left, typically colored brown. Each line represents constant potential temperature, or theta, θ, in Kelvin. When dry (unsaturated) air rises or sinks, its pressure changes. The ideal gas law states that its temperature must change. If the parcel is adiabatic, not receiving or losing heat externally, the rate of cooling or warming is a fixed, known quantity and can be defined by the dry adiabats. On the sounding, the parcel "locks" itself to a dry adiabat as it rises or falls. Thus even though its temperature changes, its potential temperature remains constant. Potential temperature is equivalent to temperature at 1000 mb, so when a parcel with a potential temperature of 290 K (following the 290 K dry adiabat) sinks to 1000 mb, its actual temperature as measured by a thermometer will be 16.9 deg C, or 290 Kelvin.

A parcel will follow the dry adiabat when it is not saturated and no external heat is added or removed.

3.4. MOIST ADIABAT GRID LINES. Moist adiabat grid lines stand nearly upright but curve to the left near the top of the chart. They are usually colored green. Each line represents constant equivalent potential temperature in Kelvin. Equivalent potential temperature is also known as theta-e, or θ_e, and like potential temperature it is calibrated to the 1000 mb level. The reason moist adiabats are on the skew T chart is to account for the release of latent heat. If a parcel is saturated, continued lift of the parcel releases latent heat into the parcel. This changes its rate of cooling with height. As a result, we cannot use dry adiabat lines. The parcel must "lock" itself to the moist adiabats and conserve its equivalent potential temperature.

A parcel will follow the moist adiabat when it is saturated. Relative humidity decreases in a sinking parcel, however, so a sinking parcel follows the dry adiabat once its relative humidity is no longer 100%.

3.5. MIXING RATIO GRID LINES. Finally, the sounding contains a set of lines showing mixing ratio. These grid lines are straight and slope up and to the right. They are usually green or blue and are dashed. These are simply reference lines which indicate the specific humidity of a parcel, thus they can be ignored except when we want to refer to them. Each line represents constant mixing ratio, w, or saturation mixing ratio, w_s.

A parcel does not follow the mixing ratio lines, but can be measured against them.

4. Lifting a parcel

An important part of working with a sounding is learning how to construct parcels which represent a cloud or thunderstorm and it vertically through the atmosphere using the environmental sounding. A parcel can be lifted any location on the sounding deemed representative. Its starting point is usually constructed from the surface level, from an elevated level, or from a layer.

4.1. LIFTING THE PARCEL. Three key principles are important in parcel lift:

(1) A parcel will lift dry-adiabatically until it cools to its dewpoint and saturates, at which point its lift will be moist-adiabatic;

(2) The humidity of a parcel is expressed in terms of mixing ratio (w), obtained from parcel temperature, and saturation mixing ratio (w_s), obtained from the parcel dewpoint. The simple ratio w_s / w yields the relative humidity; and

(3) Saturation mixing ratio is the only humidity property which remains constant when a parcel rises or sinks.

What this means is when a parcel is lifted, its mixing ratio falls as adiabatic cooling progresses and its temperature decreases. But the amount of moisture it contains remains the same, as measured by its saturation mixing ratio which is conserved. Therefore while the parcel mixing ratio is greater than its saturation mixing ratio, the parcel will cool at the dry adiabatic rate. When the two quantities are equivalent, the parcel cools at the moist adiabatic rate.

Simplifying this further, the analyst reviewing a sounding simply draws two dots on the sounding, one representing the parcel's

Predicting dry downbursts
An LCL height of 3 km or greater is usually associated with dry downburst events

starting dewpoint and the other representing its starting temperature. A line is drawn upward from the dewpoint along the mixing ratio parallels, which is used as a guide. From the starting temperature, a line is drawn upward along the dry adiabat until it intersects this guide line. This segment represents the parcel's temperature as it rises unsaturated. From this point, the parcel is lifted to the top of the chart along the moist adiabat. This segment represents the parcel's lift under saturated conditions.

4.2. LIFTED CONDENSATION LEVEL (LCL). The level at which the lifted parcel ceases rising dry adiabatically and begins rising moist adiabatically represents its lifted condensation level. At this level, the parcel's mixing ratio (w) equals the saturation mixing ratio (w_s), and the relative humidity w_s/w equals 1, or 100%. Assuming that lift is occurring, this marks the level where the bottom of a cloud base exists.

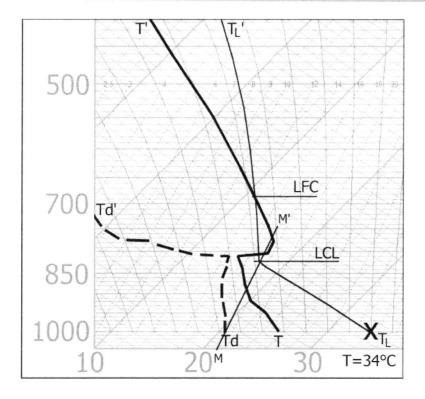

Figure 7-3. Basic parcel lift diagrammed. We assume a radiosonde launch is made and it reports a profile of temperature and dewpoint, plotted here respectively as T-T' and Td-Td'. These lines are known as the environmental profile. Note that this indicates a surface temperature of 26°C and dewpoint of 20°C. In this example we will lift a parcel which has a temperature of 34°C and a dewpoint of 20°C; thus the parcel has heated but its moisture content has not changed. The resulting lift is shown by line T_L-T_L'. Note that it rises dry-adiabatically until it reaches line M-M', where the mixing ratio is the same as that as the surface dewpoint. The parcel is now saturated and has reached its LCL (lifted condensation level). From there on it lifts wet adiabatically. When (or if) it ceases to be colder than the environmental air, it has reached its LFC (level of free convection) and will rise buoyantly. The line T_L-T_L' indicates the actual temperature of the lifted parcel and can be used to assess updraft characteristics within a storm. Line M-M' is just imaginary and is for comparing mixing ratios. Forecasters can also elect to make T_L equivalent to T, but it does not consider heating or other air mass changes and considers the stability at observation time. *(Tim Vasquez)*

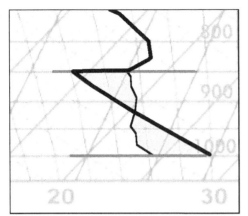

Figure 7-4a. Selecting a mean layer dry adiabat. Given a specific layer, 150 mb deep in this case, a dry adiabat is chosen, interpolating between lines if necessary, which yields equal areas at top and bottom when polygons are created bounded by the chosen dry adiabat, the environmental temperature, and the top and bottom of the layer. After the layer is thoroughly mixed, the dry adiabat actually becomes the new environmental temperature line, shown by the thick line here. Where it intersects the mixed mixing ratio line (see below) is the mixing condensation level (MCL) and it is wet adiabatic above that point.

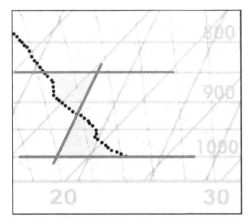

Figure 7-4b. Selecting a mean layer mixing ratio. Given a specific layer, 150 mb deep in this case, a mixing ratio isopleth is chosen, interpolating between lines if necessary, which yields equal areas at top and bottom when polygons are created bounded by the chosen mixing ratio line, the environmental moisture line, and the top and bottom of the layer. *(Tim Vasquez)*

4.3. LEVEL OF FREE CONVECTION (LFC). If the rising parcel reaches a level where it gains positive energy and can rise freely to the middle or upper troposphere, it is said to have reached its LFC. A parcel can be prevented from reaching its LFC from the existence of a capping inversion or unusually warm upper-level conditions (sometimes called "CAPE robbers").

5. Parcel methods

The most elementary way to lift a parcel is from the surface, but rarely is this representative. Therefore there are three key methods in use, each used for different purposes and weather regimes. *Instability products on the Internet occasionally do not indicate which parcel method or parcel layers are used.* Instability information from such sources should be disregarded.

5.1. SURFACE BASED (SB) PARCEL. The simplest method of lifting a parcel is to lift only surface-based parcels, assuming that all cloud material originates from the near-surface layer. In reality the atmosphere never forms updrafts entirely from a surface layer. A surface based method is also significantly unrepresentative if the air mass is different just above the surface, as with the example of a very thin layer of tropical air at a given station. In some cases the surface-based method is the only one available, and this is fine for obtaining a first guess at instability. However the forecaster should elect to use the mixed layer method or most unstable method whenever possible.

5.2. MIXED LAYER (ML) PARCEL. When solar heating is occurring, the atmosphere will mix through deeper and deeper layers, blending and equalizing the potential temperature and specific humidity throughout the layer. Large portions of this blended layer are ingested by cumulus and cumulonimbus updrafts, so a mixed layer parcel method offers a very accurate representation of updraft air. Research has also shown that mixed layer parcels are much more representative for predicting cumulus cloud base height than surface based parcels (Craven and Jewell 2002).

To use the mixed-layer method, a layer is always selected by defining its top and bottom. It is usually defined as a depth in millibars, with its base usually at the earth's surface. A depth of 100 to 150 mb is preferred, but the analyst may elect to choose their own layer based on details in the sounding.

Once a layer has been defined, the goal is to find a *mean mixing ratio* for the dewpoint trace and a *mean dry adiabat* for the temperature trace. See the illustration for techniques on how to do this. Once a mean mixing ratio and mean dry adiabat line have been found, these mean values replace the existing environmental moisture and temperature trace form the basis for a parcel to be lifted. Where they intersect, this defines a mixing condensation level (MCL), and above that the parcel rises wet adiabatically.

5.3. MOST UNSTABLE (MU) PARCEL. The most unstable method seeks to identify the level in the lowest 300 mb of the atmosphere that produces the most unstable parcel. This is the level that produces the warmest parcel relative to any arbitrary level aloft, or in more technical terms, produces the highest theta-e. The lowest 300 mb of the atmosphere spans a depth of about 10,000 ft, or about 3 km.

The most-unstable method should always be selected in situations where convection is expected above a stable layer, in other words, with elevated convection. This type of situation occurs poleward of frontal boundaries. The presence of colder layers near the ground will corrupt the value of SB and ML methods as parcels at these levels will not be ingested by elevated convection. Therefore the MU method is favored.

Instead of lifting a parcel with any of these methods, the forecaster may also elect to manually lift a parcel from a single elevated level or construct an elevated mixed layer from which to lift a parcel. This may be necessary if the cold layer is greater than 300 mb deep, causing many automated MU methods to fail. Upper-level parcel lifts are not found on the Internet and will require the forecaster to hand-analyze soundings or use sounding analysis tools such as SHARP or RAOB.

5.4. EFFECTIVE INFLOW LAYER. A new technique called the effective layer method (Thompson 2005) examines the sounding to identify levels containing parcels with positive buoyancy and groups them into a single layer. As many diagnostic quantities of shear and instability are based on rigid, fixed layers, the effective layer method shows promise for yielding more accurate quantities.

To compute an effective layer, two predetermined constants are needed: the minimum CAPE and maximum CINH. These are by definition 100 J·kg^{-1} and -250 J·kg^{-1}, but they may be changed by the analyst. Starting at the ground and progressing to higher levels in very fine steps, parcels are lifted from each level and evaluated for CAPE and CINH. The lowest level which meets the effective layer criteria is classified as the effective inflow base. The highest level which meets the effective layer criteria is classified as the effective inflow top. Since this method has specific threshold requirements for CAPE and CINH it is almost always done by computer methods.

6. Modification

In the United States, soundings only show a snapshot of the atmospheric conditions in the morning hours. Significant changes can occur by afternoon. Instability calculations performed on the morning sounding only show instability at the time of the sounding and not what will be present later. One of the marks of a seasoned forecaster is how well they can modify the sounding to account for changes in the atmosphere at convection time.

A modified sounding will change the environmental temperature and dewpoint trace. This, in turn, will affect the characteristics of the selected parcel.

6.1. ADVECTION. One of the factors that causes changes in temperature and dewpoint in the column is the advection of air with different properties from various locations at different levels. A change of two Celsius degrees, for example, can make all the difference in eliminating a cap. However because of the tendency for air to flow isentropically rather than horizontally, advection cannot be estimated simply by looking for cold air advection or warm air advection on constant pressure charts.

Model output is perhaps the best tool for estimating changes in the sounding due to advection, particularly in the middle and upper levels. It is possible to use a model forecast sounding directly, however these forecasts rarely have good correlation with actual soundings.

6.2. POTENTIAL INSTABILITY. Cold, dry conditions aloft and warm, moist conditions in the low levels create an atmosphere with potential instability. This is realized when a source of dynamic lift occurs, such as the approach of a strong upper-level disturbance. When the column is lifted, the upper portions rise dry adiabatically and cool quickly, while the lower portions rise moist adiabatically after condensing and cool slowly due to the release of latent heat. This strong cooling aloft and weak cooling in the low levels steepens the lapse rates.

6.3. CONVECTIVE CONDENSATION LEVEL (CCL). The CCL is the level at which a parcel will become saturated *when enough solar heating is added to produce deep convection*. This is calculated by performing a crude bit of sounding modification. The procedure is to locate a representative low-level moisture parcel, find the mixing ratio line that crosses this parcel, then locate where this line in turn crosses the environmental temperature trace. The level at this spot is the CCL. The environmental temperature trace is then erased below this point and reconstructed downward from the CCL along the dry adiabat to show the hypothetical sounding at the time that deep convection occurs. Where this meets the surface level, the *convective temperature (CT)* is obtained.

The technique must be used with caution, though, as it goes on the assumption that enough heating has occurred to break the cap. It has no ability to predict whether this will actually occur. Therefore forecasters using this method must compare the forecast maximum temperatures to the convective temperature, and adjust the modified sounding accordingly. It also prescribes using a surface parcel, thus giving a surface-based parcel method. It is often more prudent to use a mixed-layer or effective inflow parcel.

7. Sounding proximity

The key to understanding soundings is to be aware that the vast majority of them are used as environmental soundings — that is, the sounding shows the profile of the environment that a storm is growing into, *not the profile within the storm cloud*. Most of the work that a forecaster does with a sounding is meant to simulate the ascent and descent of parcels within this environmental air and determine whether these ascents and descents are possible and how strong they will be. This, in turn, indicates what kind of weather will occur.

7.1. PROXIMITY SOUNDING. An environmental sounding that is considered to be highly representative of a specific storm environment is described as a proximity sounding. It has no fixed definition, but the basic definition (Darkow 1969) involves a sounding that is taken within 50 miles and 105 minutes of a storm and is representative of the air mass that nurtured the storm. Some definitions expand this out to 3 hours and 120 miles, while storm-scale research uses much narrower criteria, sometimes requiring launch immediately next to a storm cloud. Proximity soundings are somewhat rare.

7.2. PARCEL SOUNDING. A parcel sounding samples the actual inflow of air rising into a deep convective cloud. They are quite rare and are never performed intentionally except in field research projects. Generally these soundings show a very deep, saturated profile parallel with moist adiabats. A parcel sounding might reveal information about the potential temperature and specific humidity of the inflow and updraft into the cloud. In most routine radiosonde networks this is considered representative of the inside of a single convective cloud, not the environment, so it is generally considered to be convectively contaminated and ignored.

8. Sounding characteristics

Now that we have a basic understanding of how to read the sounding and lift parcels on it, we can now cover some of the more qualitative elements of interpretation. Without any special techniques or calculations we can simply glance at the sounding and find information on lapse rate, instability, caps, the potential for convective weather, and much more.

8.1. LAPSE RATE. Perhaps the simplest measurement of instability is lapse rate. Visually, this looks like a temperature segment on the sounding which leans to the left. Quantitatively, lapse rate equals $\gamma = -dT/dz$, in other words the decrease in temperature over a given vertical interval, usually given in Celsius degrees per kilometer. One way that lapse rate is quantified is the Vertical Totals (VT) index, which measures the difference between the 850 mb and 500 mb level in Celsius degrees. This preselected layer is *usually* between the low-level tropical layer and the lower reaches of the upper troposphere, making it a useful measure of lapse rate for severe thunderstorm forecasting.

8.2. POSITIVE AREA. If the temperature and moisture properties of a given parcel have been properly selected, its ascent on the sounding can be plotted showing its temperature properties as it rises. The "area" on the sounding between the temperature profiles of the ascending parcel and the environment yields a measure of energy. If the parcel ascent is warmer than the environmental temperature, then the area is considered to have positive energy, or positive area.

8.3. DRY ENTRAINMENT. Even with favorable CAPE and CAPE densities, an updraft might entrain dry air, which slows its ascent. This may happen when a narrow updraft rises through the cap, allowing a "film" of dry air from the surrounding environment to be drawn up into the updraft. Wide updrafts reduce the effect of dry air entrainment, limiting it to the outer fringes of the updraft. Dry air entrainment will be less than expected if a tower grows in an area of humidified air, such as where other towers have been growing or in an area of large-scale lift, such as near a persistent boundary.

8.4. CAP. The cap, or "capping inversion", is a layer of warm air above the surface. It acts as an inversion to delay the development of deep convection. When development does occur, it tends to oc-

The use of indices

In my experience, many forecasters, implicitly or not, are seeking a "magic bullet" when they offer up yet another combined variable or index for consideration. If forecasting were so simple as to be capable of being done effectively using some single variable or combination of variables, then the need for human forecasters effectively vanishes.

CHARLES DOSWELL, 2006

cur only in areas of strong mesoscale forcing and later in the day when solar heating has resulted in much warmer, buoyant parcels.

The configuration of plateaus and mountain areas in North America lends itself to development of the cap on the Great Plains. Other areas of the world where storms are common, such as the Atlantic seaboard, Europe, and Pacific Rim, rarely encounter capping situations. Without a cap, storms tend to break out early just as ingredients come together, and occur across a wide area and with relatively weak intensity rather than late in the day in strong, isolated modes.

How does a cap work? As a parcel of air rises into this cap, the air around it becomes relatively warmer than the parcel. This makes the rising parcel less buoyant, and it loses momentum, ultimately sinking if it is still surrounded by relatively warm air. The only way the parcel can break the cap is to increase its theta-e temperature by heating further or increasing its specific humidity.

If a parcel can break through the cap and reach the large positive area, then it is said to have reached its level of free convection, or LFC. It reaches its LFC by either becoming warmer than the surrounding air, by forced motion through the cap, or by weakening of the cap through synoptic-scale lift.

Since the tropical air mass beneath the cap is confined and unable to disperse into the free atmosphere, theta-e of the surface layer has a tendency to build with the combined effects of heating and moisture advection. When the cap breaks, parcels have the potential to be extremely buoyant because of the increase in theta-e, and this energy is realized when they reach the steep lapse rates aloft.

8.5. ELEVATED MIXED LAYER (EML). The elevated mixed layer is a deep "plume" of warm, dry air that advects through the mid-levels of the troposphere. Since it is warm, it tends to produces an thermal inversion at its base. This inversion, coincidentally, is the cap. The EML is rooted in desert regions, in particular, the northern Mexican interior and the southwest United States. In its source regions it is characterized by nearly adiabatic lapse rates (i.e. very cold air aloft and very warm air at the surface). The dryline marks where this desert air mass meets tropical moisture. The EML lies atop the tropical moisture and can often be identified eastward into the eastern United States.

Convective "lift"

Forecasters seem to have a difficult time conceptualizing parcel lift. For parcel theory to work, single parcels need to be lifted relative to surrounding environmental air that is not lifted. Hence, verbiage like "...the cold front will lift parcels to their LFC..." or "...parcel lift related to jet-streak dynamics..." is incorrect. Any synoptic scale or high-end mesoscale lifting mechanisms do create an environment more favorable for convection by regionally steepening lapse rates (for the moisture stratification typical in the Plains) and lowering the LFC closer to the surface, thus making it "easier" for parcels to be lifted to a reasonable elevation.

JOHN MONTEVERDI, 2009
personal communication

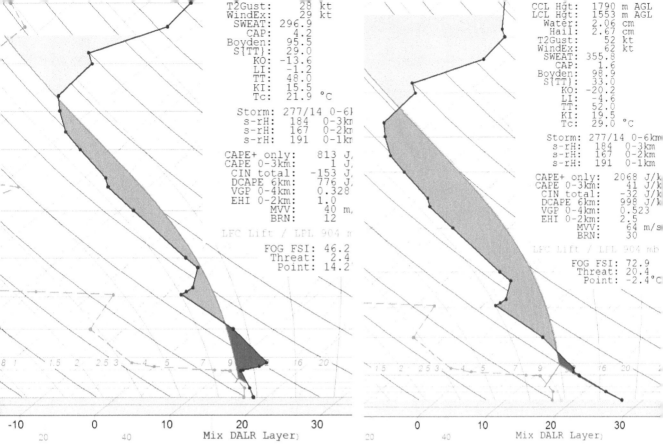

Figure 7-4a. Unmodified sounding for 3 May 1999 1200Z, Norman, Oklahoma, date of the deadly Oklahoma tornado outbreak. The software has been configured to show a mixed layer parcel (lowest 125 mb) and using this, shade the CIN (bottom), CAPE (middle), and overshoot (top) areas. Note that there has been no modification at all for surface heating. The CIN is a very stout -153 J·kg⁻¹, and a quick glance at the diagram shows that it would take a surface temperature of nearly 30 deg C to break the cap. (Tim Vasquez / graphics with RAOB <raob.com>)

Figure 7-4b. Modified sounding for 3 May 1999 1200Z, Norman, Oklahoma. The temperature plots have been manually modified by dragging them to the forecast maximum temperature (28 deg C) and creating a dry adiabatic lapse rate above that to reconnect with the sounding. This is a best guess at the afternoon profile, not accounting for any upper-level changes. The cap is now breakable and significant instability is found above the LFC. The modified mixing ratio uses the software's precomputed mean lower 125 mb value of 11 g·kg⁻¹.

8.6. ELEVATED CONVECTION. It is quite common for convection to exist above inversions, including caps and frontal inversions (Colman 1990). This type of convection is said to be rooted above the boundary layer, or above the surface. Elevated convection occurs if sufficient moisture and instability is above the inversion.

The air within and above capping inversions and subsidence inversions are usually too dry to support elevated convection. However moisture which is trapped within the upper fringes of the cap-

The role of shear

The 1940s-era Thunderstorm Project established two key findings about wind shear: that it disrupts growing cumulus towers but correlates with long-lived mature cells. By 1960 it was proposed that this shear advects the precipitation cascade downwind, keeping it from interfering with the updraft. As an analogy, shooting a garden hose directly at the zenith will get the sprayer wet unless there is strong wind blowing, which will carry the droplets away. This strong wind aloft prolongs the life of the updraft and produces what is called a long-lived storm. In meteorology this type of shear is often expressed by the 0 to 6 km AGL bulk shear.

As the Thunderstorm Project discovered, strong shear shreds not only small cumulus towers apart but also large updrafts. This forms what is now known as "turkey towers", towering cumulus clouds that have long-necked appearances and are eventually pulled apart. Close inspection of cumuliform towers also shows that they lean, much like the shape of the Tower of Pisa, tilting with height in the downshear direction.

ping inversion can occasionally mix out into the steep lapse rates above, producing altocumulus castellanus.

Elevated convection can also occur in dry air masses, such as behind the dryline where weak moisture condenses at very high condensation levels, producing anything from high-based cumuliform clouds to castellanus-like towers and virga. However this type of convection, since it is rooted at the surface but is extremely high, is better referred to as *high-based convection*.

A prolific source of elevated convection is a frontal inversion, which is usually associated with humid layers within and above the inversion. Storm structure is usually hidden by extensive stratus or stratocumulus decks occupying the frontal transition zone or the cold air mass itself.

Elevated severe storms are not coupled to the cold air mass underneath and they can exist even when the cold air mass is extremely frigid. On 22 February 1975 several damaging F2 tornadoes moved through southwest Oklahoma while surface temperatures were near 40 deg F (5 deg C) with north winds! If the polar air mass contains subfreezing temperatures and is sufficiently deep, sleet or freezing rain may occur, caused by liquid precipitation that freezes near or upon the ground. This wintry surface weather is completely independent of any processes taking place in the storm cloud itself. In the vast majority of cases, however, severe weather is limited to hail and high wind. The cold layer near the surface, though not buoyant enough to make up the storm inflow layer, is believed to hinder tornadogenesis processes in some way.

It should be noted that elevated storms can also be the result of convection which has formed within the warm sector but has crossed a frontal boundary onto cold air. The storm tends to become elevated and the severe weather threat diminishes as the storm crosses over deeper and deeper layers underneath.

8.7. CAPE ROBBER. Though the mid-level capping inversion is recognized as a problem with storm initiation, abnormal warmth can instead occur in the upper troposphere. Such warmth shrinks or removes the top of the positive energy area. This feature is the so-called "CAPE robber". It was a significant problem during the 1998 Great Plains storm season, and has occasionally appeared in other years.

STABILITY & SHEAR   141

A CAPE robber scenario tends to affect a synoptic-scale region and is typically caused by strong subsidence in the upper troposphere.

8.8. DOWNDRAFT CONVECTIVE AVAILABILITY OF POTENTIAL ENERGY (DCAPE). The DCAPE measurement is similar in principle to CAPE but is slightly more complicated and is subject to dependencies on which levels are selected. First the most representative wet-bulb potential temperature (theta-w) for the updraft is obtained; this will generally be the moist adiabat that the saturated, rising parcel follows.

Again, DCAPE is dependent on which level the forecaster selects. DCAPE does not account for precipitation loading on the parcel. One model simulation in the late 1990s suggested that DCAPE was a poor indicator of downburst strength.

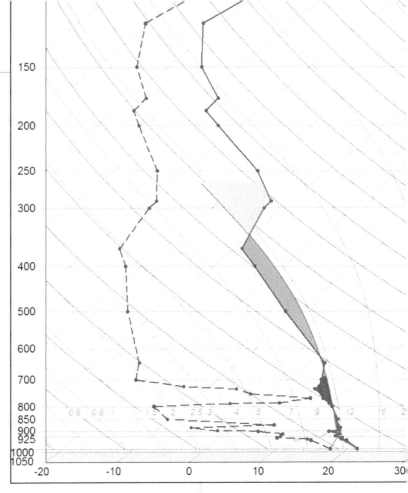

Figure 7-5. Cape robber in Texas in April 1998. Even though the tropopause is near the top of the diagram, a warm layer exists down to about 375 mb, forming an inversion. The CAPE robber causes storm tops to remain mostly below 350 mb (27,000 ft) and prevents them from realizing their full buoyancy in the free atmosphere. *(Tim Vasquez / graphics with RAOB <raob.com>)*

9. Thermodynamic diagnostics

Properties of the sounding can be summarized by the use of numerical quantities. These must of course be used with caution, as all diagnostics are based on assumptions.

9.1. SHOWALTER STABILITY INDEX (SI, SSI). The SI is one of the oldest diagnostic variables in existence, having been developed in the 1940s by Showalter. It is defined as:

$SI = T_{E500} - T_{L500}$

where T_{E500} is the environmental temperature at 500 mb and T_{L500} is the final temperature of a parcel lifted from the 850 mb

> **Attention to detail**
>
> Under deadlines and distractions, even experienced forecasters are tempted to take the easy way out: sacrifice time-consuming hand-analysis of surface and upper air charts for the addictive, quick-fix drug of objectively analyzed fields. Such tools do have their place; indeed, they might be good enough much of the time.
>
> The most unusual, extreme and deadly events, however, require more than "good enough" -- they demand *excellence*. They often involve small processes not well-handled by models, depend on subtleties poorly depicted in machine-analyzed fields, and require the deepest level of analytic understanding and skill possible. Detailed and thorough hand analysis still has no substitute as a critical tool for true *insight* into the current and recent state of the atmosphere. A correctly analyzed nuance can make the difference between forecasting and missing a career event!
>
> ROGER EDWARDS, 2009
> personal communication

level to 500 mb. In short, this compares the temperature of a parcel with its environment.

9.2. LIFTED INDEX (LI). The Lifted Index was developed by SELS forecaster Joseph Galway in the early 1950s and published in 1956. It is very similar to the SI in comparing a parcel lifted to 500 mb, except instead of using a parcel starting at the 850 mb level, it uses a parcel starting at the earth's surface using the forecast temperature and dewpoint.

$$LI = T_{E500} - T_{L500}$$

9.3. CONVECTIVE AVAILABILITY OF POTENTIAL ENERGY (CAPE). In the 1990s, CAPE became the favored measure of instability by severe storm forecasters.

$$CAPE = \int (\alpha_{lp} - \alpha)dp$$

where $\int$ indicates an integral from the LFC to the EL.

CAPE by itself is very effective at distinguishing non-thunderstorm environments from thunderstorm environments. Its skill in distinguishing the possibility of supercells from non-supercells is marginal, however, with 500-1000 being a lower end MLCAPE value for supercells. For forecasting tornadoes it is not reliable. Combined parameters have been found to work much better for predicting these occurrences than CAPE alone.

Physical calculations show that it's theoretically possible to determine updraft strength from CAPE by the relation $V=(2 \times CAPE)^{0.5}$ where V is the speed in m·s^{-1}. In reality, this is slowed by about 50% due to precipitation loading and dry air entrainment, and in rotating storms it can be enhanced by vertical pressure gradients. CAPE is very sensitive to which low-level parcel is selected. Since it is impossible to predict exactly which layer will be ingested by the updraft, thus there is no truly accurate CAPE number. CAPE is expressed by the type of parcel selected, such as MLCAPE (mean layer CAPE), SBCAPE (surface based CAPE), or MUCAPE (most unstable CAPE).

Finally, CAPE is not a prerequisite for severe storms. Mini-supercells, some of which are even tornadic, can develop in regions of very low CAPE, on the order of only 300 J·kg^{-1} (Markowski and Straka 2000).

9.4. CONVECTIVE INHIBITION (CINH). The concept of convective inhibition is almost identical to that of CAPE, however instead of measuring the large positive area, the negative area beneath it is measured. This is done, as with CAPE, by measuring the area between the environment and parcel lift and obtaining a result in J·kg^{-1}. Values of 10 J·kg^{-1} or below are considered weak, 10 to 40 moderate, and above 40 strong.

Unfortunately, as with CAPE, CINH is strongly sensitive to which parcel is selected for the parcel lift. Furthermore, the difference between boom or bust is only on the order of tens of J·kg^{-1}, which equates to only a few boxes on a detailed SKEW-T! Internet charts that show CINH can be unrepresentative unless there is a complete understanding of how the parcels were selected and whether they are appropriate for the given weather situation. Furthermore, the numbers also depend on high accuracy of mid-level temperature profiles. Vertical pressure gradient forces in a developing updraft can couple with buoyancy to overcome the cap, a process which is not well-understood and is not predictable. Finally, since CINH is strongly sensitive to temperature and moisture, CINH diagnostic fields may produce unusable output in regions of high thermodynamic gradients, such as near fronts and drylines.

For these reasons, the use of convective inhibition maps is strongly discouraged without a detailed understanding of how the lifted parcels are selected and whether they are appropriate for the forecast situation. In most cases, cap evaluation is an exercise best left to human scrutiny of station soundings.

9.5. CAPE DENSITY. Some researchers have explored the relationship between CAPE and its depth. Short, fat positive areas correspond to high CAPE density, while tall skinny positive areas equate to low CAPE density. It has been proposed that tornadic storms strongly favor atmospheres with high CAPE density, as this might allow for stronger acceleration of the updraft. This has been confirmed by numerical modelling simulations (Wicker and Cantrell 1996) which developed rotating storms with high CAPE density but values of only 600 J·kg^{-1}.

One possible reason is that low CAPE density values are associated with slow acceleration of the updraft, which allows more time

What u and v mean
Occasionally forecasters may encounter research papers describing winds in terms of u and v coordinates. Here's what they mean.

Positive values of u indicate a wind component blowing from west to east.

Negative values of u indicate a wind component blowing from east to west.

Positive values of v indicate a wind component blowing from the south to the north.

Negative values of v indicate a wind component blowing north to south.

for precipitation to form within the updraft, which enhances precipitation loading and slows the updraft. The opposite is true with high CAPE density. Therefore updraft acceleration can be expected to be stronger in high CAPE density environments.

A measure called normalized CAPE (NCAPE) was developed (Blanchard 1998) which simply divides CAPE by its depth in meters. A value of 0.1 is associated with low CAPE density and thin, skinny positive areas with weak parcel accelerations, while values of 0.4 are associated with high CAPE density, fat positive areas, and large parcel accelerations. Note that this yields no actual information about CAPE or energy itself; for example, high values of NCAPE can exist with no significant instability.

9.6. LIFTED CONDENSATION LEVEL (LCL). LCL is a proxy for relative humidity within a layer. A high LCL height is indicative of an environment with low relative humidities, which produces strong, cold thunderstorm outflow and produces downdraft-dominant modes. A low LCL height favors weak outflow and updraft-dominant modes. MLLCL values of less than 1000 m are associated with moderate to strong tornado environments.

10. The hodograph

A hodograph is an essential tool for the severe storm forecaster and must be understood. Its sole purpose is to visually display the shear that is present in the atmosphere. Strong shear is directly correlated to storm organization. It's also possible to make many inferences about storm behavior. Like a sounding, a hodograph shows upper air conditions for one station at a given point in time. It is constructed entirely from the wind data.

10.1. CONSTRUCTION. The hodograph presents a radial graph of *ground-relative* wind direction (azimuth from the center) and speed (distance from the center). So if a radiosonde site reports calm winds at all levels, the entire hodograph consists of a single dot at the center.

To construct a hodograph plot, the forecaster marks wind data on the graph from the ground to the tropopause at a given station. For example, if the 0 km (surface) wind is blowing *toward* the

STABILITY & SHEAR 145

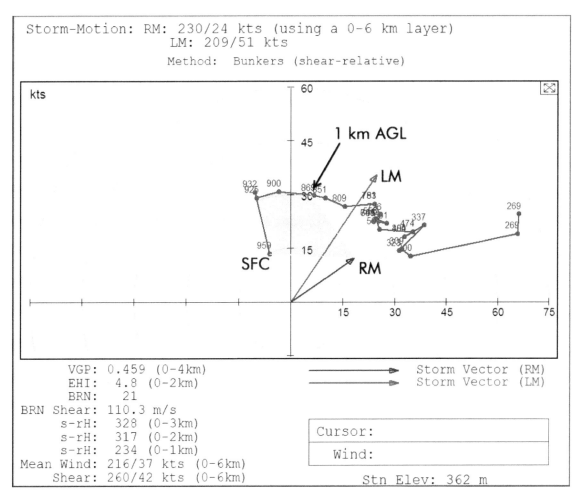

southeast at 20 kt, a point is plotted on the 135° radial and at the 20 kt ring. It must be remembered that hodograph plots represent physical wind forces, so we must always plot winds in the direction *toward* which they are blowing. Since wind reports are universally expressed as the direction *from* which they blow, even in radiosonde reports, this is a common source of confusion for beginners.

Once all of the points are plotted, a continuous, unbroken line is drawn connecting all the points from the ground upward. It is vital that all of the points are labelled with respect to height. It is common for some websites and software programs to display hodographs without any annotations. This is useless for forecasters since it gives no indication which atmospheric layers are experiencing the indicated shear and how deep these layers are.

Figure 7-6. Hodograph for 3 May 1999 in Norman OK at 7 pm CDT. The Bunkers method storm motion vectors are plotted. Note that from the perspective of the right (RM) storm motion vector, a large area is swept out in the lowest kilometer of the atmosphere (shading). The geometric area of this "swept out area" is proportional to storm-relative helicity, a key ingredient in storm rotation. *(Tim Vasquez / graphics with RAOB <raob.com>)*

Total shear

Total shear has a correlation with storm organization. Though severe storms can form with even a modest hodograph length, a length for the layer 0 to 3 km of more than 40 kt has been conclusively shown by numerical simulations to be associated with very long-lived, organized storms.

10.2. HODOGRAPH SHAPE. The length of the hodograph line for a given layer is directly proportional to the total shear through that layer. Interestingly this line is fractal in nature: if the atmosphere is sampled in greater detail, it adds more minor jags to the line, artificially inflating its length.

If the hodograph shape is fairly straight, this indicates unidirectional shear vectors. This is associated with splitting storm modes. On the other hand, curvature of the hodograph line is very significant for the severe storm forecaster. This illustrates a property known as helicity. The curvature is described further by which direction of turn occurs with increasing height: *clockwise* or *counterclockwise*.

A curved hodograph line implies that streamwise vorticity is present in the wind field through the layer. If this layer coincides with the storm inflow, the risk is very high of mesocyclone development and deviant storm motion. In the northern hemisphere, a clockwise curve in the inflow layer is the most common mode and is associated with right movement and cyclonic storm rotation, while a counterclockwise curve is much less common and is associated with left movement and anticyclonic storm rotation.

10.3. SHEAR. Shear is a change in the wind vector with height (Glickman 2000). The expression known as *cumulative shear* or *total shear* is integrated with height, that is, calculated between two levels with very small steps and considering all changes in between. The cumulative shear within a given layer is equivalent to the length of a hodograph line in that layer, regardless of its shape. The word "shear" in severe thunderstorm forecasting without any context usually refers to this cumulative shear quantity.

The quantity *bulk shear*, on the other hand, is simply the shear vector between the top and bottom of a layer without considering any properties within the layer. This is a simplistic expression but is easy to calculate and visualize. For severe weather forecasting, 0 to 6 km AGL has long been a traditional layer for evaluating bulk shear, but other layers can be used instead.

The term *vertical shear* emphasizes that it measures changes along the vertical axis, to differentiate it from *horizontal shear* which considers winds in the horizontal plane. In severe weather

STABILITY & SHEAR

forecasting, shear quantities almost always refer to vertical shear, with the exception of Doppler radar velocity products which traditionally display horizontal shear at small scales.

10.4. DIRECTIONAL SHEAR. To the meteorological community in general, directional shear refers to a *wind vector* that changes direction with height. But to the severe storm community, and within this book, the term describes a *shear vector* that changes direction with height. This distinction is an important one for the forecaster. The winds from the surface to 850 mb may be southerly with some variations, suggesting a wind vector without much directional change, but when plotted on a hodograph, the line may show large curves with shear vectors pointing in many different directions.

If there is a lack of directional shear, it is referred to as a "unidirectional" profile. The hodograph trace tends to be straight. It is important to note that a glance at radiosonde-observed winds on a sounding, which are ground-relative, are not sufficient to assess whether unidirectional or directional shear profiles exists.

10.5. 0-6 KM BULK SHEAR. As mentioned earlier, it's possible to calculate shear by obtaining the difference of winds at two levels, or on the hodograph simply measuring the distance between the two

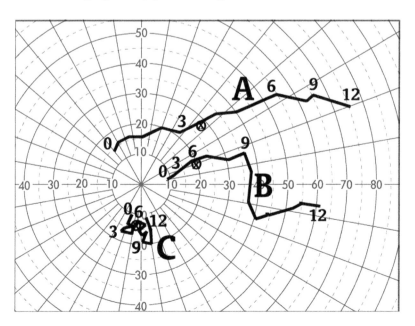

Figure 7-7. Three types of hodographs with low helicity. Hodograph "A" shows a unidirectional shear profile; an radiosonde launch would show southeast winds at the surface and southwest winds aloft, but from a storm-relative perspective all the winds are either blowing to or from the west. Hodograph "B" shows strong curvature, but not in the critical lowest 3 km; all the curvature is high up in the storm at 6 to 10 km. This illustrates the danger of interpreting hodographs which use unlabelled points; unfortunately a very common occurrence with Internet sites. Hodograph "C" shows very weak storm-relative flow, caused in this case by winds throughout the troposphere blowing moderately from the north. *(Tim Vasquez)*

points representing these levels. The level 0 to 6 km AGL is commonly used, yielding 0-6 km bulk shear. It is a predictor of separation between the updraft and downdraft, indicating storm longevity and severity, but the parameter has been found to be useless for differentiating tornadic environments from nontornadic ones. Still it can be useful for identifying supercell environments. A recent study (Houston et al 2008) evaluated many different layers and found that the 0-5 km AGL bulk shear provided the best results, with values of 26 kt or above strongly associated with supercells and values below 26 kt strongly associated with non-supercells.

10.6. BULK RICHARDSON NUMBER (BRN) SHEAR. The BRN shear value is essentially the denominator of the Bulk Richardson Number. It is similar in principle to 0-6 km bulk shear but uses different layers. BRN shear is defined as $U = (U_1 - U_2)$, where U_1 is the mean wind in the 0-6 km layer, and U_2 is the mean wind in the 0-0.5 km layer, with all velocities in m·s^{-1} and the mean wind ideally being density weighted (i.e. biased somewhat toward values at lower heights). The BRN shear can be constructed on a hodograph by mentally extracting the 0 to 6 km hodograph line and identifying a point that this line would "balance" upon, and doing the same for the 0 to 0.5 km hodograph line. The length of the vector between these two points indicates the BRN shear, and can be measured using the radial grid of the hodograph as a scale. Multiplying a figure in knots by 0.514 will give the required units of m·s^{-1}.

BRN shear is rather good at discriminating between tornadic/supercell environments and non-supercell environments, but is poor at distinguishing between types of supercell environments. Values of below 20 m·s^{-1} indicate non-supercell storms, while those above 25 m·s^{-1} indicate supercell storms. Those above 35 m·s^{-1} tend to have well-defined supercell characteristics.

10.7. EFFECTIVE BULK SHEAR. The effective bulk shear (Thompson et al 2005) replaces the traditional, fixed 0-6 km bulk shear quantity with data from a more meaningful layer for a given situation. This properly samples the inflow and anvil, providing much better estimates for unusual situations like mini-supercells, cold core convection, and elevated storms. The effective bulk shear is

STABILITY & SHEAR 149

the wind vector between the *most unstable parcel level* and the *equilibrium level*.

10.8. 0-1 KM BULK SHEAR. The bulk shear between 0-1 km AGL has recently shown promise for discriminating tornadic from non-tornadic supercells. In a sense, it is a crude proxy for storm-relative helicity (to be discussed shortly) since hodograph curvature elongates the hodograph line. One study found that 0-1 km AGL shear of 20 kt or more was associated with a significant tornado risk.

10.9. STORM MOTION. Many important quantities on the hodograph depend on an accurate estimate of storm motion. When storms are underway this can be observed directly by measuring movement on radar. In other cases it must be calculated by examining the environmental winds and estimating a "steering flow" that moves the storm. Once storm motion is determined, the movement of deviant cells must be estimated. Basic storm motion is easy to calculate, but deviant motion is more difficult.

On the hodograph, basic storm motion consists of a point on the hodograph corresponding the mass-weighted mean position of the entire tropospheric hodograph line (mass-weighted means that all levels are treated with equal weight), though in practice the 0 to 6 km hodograph segment has been found to work much better. This can often be "eyeballed", but computer based tools can calculate this automatically and with greater accuracy. With shallower storms, 0-6 km mean winds may be unrepresentative and a shallower steering layer is prescribed, with the opposite suggested in a high-topped storm situation.

A method that was used often during the 1980s and 1990s was the 30R75 technique (after Maddox 1976), which prescribes that a right-mover will move 30 degrees to the right of the storm motion (relative to the ground) and at 75% of its speed.

A new technique called the *Bunkers method*, also known as the *internal dynamics (ID)* method (Bunkers et al 2000), has rapidly gained popularity. Its construction requires that a line is drawn from the mean 0-0.5 km AGL portion of the hodograph line to the mean of the 5.5 to 6 km portion of the line. A non-severe storm motion vector is calculated using the mean 0-6 km wind (i.e. a point at the correct location on the hodograph that would cause the 0-6 km hodograph segment, if it were an imaginary object, to "balance" upon it). A secondary line is drawn orthogonal to this first line which intercepts the predetermined storm motion point. Then two vectors, each of 7.5 m·s^{-1} (15 kt) in magnitude, are drawn from the storm motion point each way along the secondary line. The point at the tips of each of these vectors forms the result: the right mover and left mover motion. The right mover motion is always on the right-hand side when looking downshear.

A further improvement is the *effective Bunkers* or *effective internal dynamics (ID)* technique. It avoids rigid use of the 0-6 km layer and instead uses a cross section between the effective inflow base (see "effective inflow layer") and the midpoint between the effective inflow base and the equilibrium level. However some preliminary studies (Thompson et al 2005) indicate no improvement in skill over the original Bunkers method.

10.10. VORTICITY. Crosswise vorticity is the component of vorticity along an axis perpendicular to the mean shear. An analogy of this is throwing a football so that it stays vertical and spins end over end, with one end coming toward you and another going away. When storm motion lies on the hodograph, all of the vorticity is crosswise.

Streamwise vorticity is the component of vorticity along an axis parallel to the mean shear. An analogy is throwing a football so that it stays vertical but spins end over end, with the spin direction perpendicular to your throw. When the storm movement is not on the hodograph trace, there is streamwise vorticity. A high amount of streamwise vorticity in the 0-1 km AGL layer is considered to be one of the best predictors of tornadic activity.

10.11. STORM-RELATIVE HELICITY (SRH). In terms of a hodograph, storm-relative helicity can best be described as the geometric area between (1) a single point at the tip of the storm motion vector and (2) the line segment on the hodograph representing the inflow layer. The greater of an area that is "swept out", the larger the SRH. The quantity is also known as storm-relative environmental helicity (SREH), which is discouraged as it presents ambiguity of what "environmental" refers to.

STABILITY & SHEAR

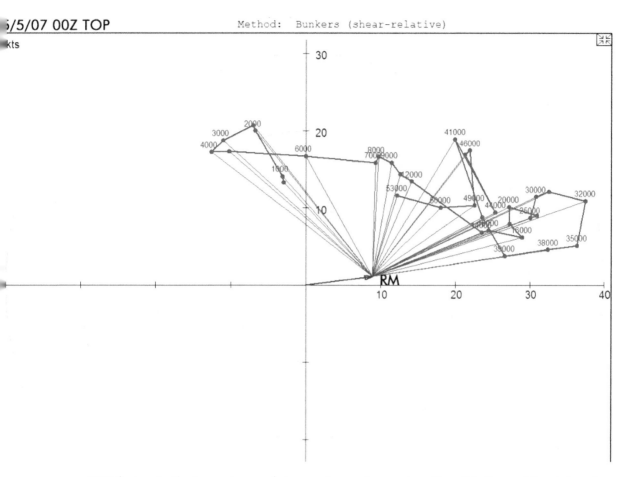

SRH factors in the importance of *streamwise vorticity* on a rotating storm (Davies-Jones 1990). Research has shown that SRH is a very strong predictor for tornadic environments, with one study of 6000 soundings showing values of above 100 $m^2 \cdot s^{-2}$ being associated with 75% of tornadic events and those below being associated with 75% of nontornadic events (Rasmussen and Blanchard 1998). For strong tornadoes, a mean SRH of about 360 $m^2 \cdot s^{-2}$ was found in one study (Johns et al. 1990, Davies-Jones et al. 1990) though another study found lower numbers, near 200 $m^2 \cdot s^{-2}$ (Kerr and Darkow 1996).

SRH has often been used by itself as a diagnostic variable to assess the potential for supercells and tornadoes. The definition of SRH holds that values of above

Figure 7-8. Storm-relative winds as drawn from the perspective of the right-motion vector. The numerous lines show storm-relative wind at different levels. If we assume that we are riding with the storm in a hot air balloon (which managed to exactly match the speed of the storm) we would notice the lower part of the storm moving from southeast to northwest and the upper parts from southwest to northeast. The "swept out" area is pretty small from 0-1 km (1000-4000 ft AGL here) but large for 0-3 km (1000-10000 ft AGL). This diagram shows the Topeka sounding on the evening of May 4, 2007, the Greensburg F5 day. Topeka is about 250 miles away but this still gives some idea of the shear profiles in Kansas that evening.

Predicting tornadoes

Predictions of LCL height and 0-1 km shear/SRH have been very reliable (not perfect!) indicators of tornado potential on the mesoscale and outlook time scales, given the right storm mode. Effective bulk shear helps to separate supercell from non-supercell environments, even though there is some overlap. Combinations of these variables with CAPE, such as SCP and STP, can signal areas to watch, including "sneak attack" potential not currently in the forecast.

But remember: If one variable is barely below thresholds, the entire index fails to activate. For example, a slight increase in CAPE from one hour to the next, in an area of already-intense shear and low LCL, suddenly can introduce big areas of alarming-looking STP "out of nowhere". In fact, the potential already existed and has risen only a little. Forecasters using indices must know the status of each component variable!

ROGER EDWARDS, 2009
personal communication

---

150 $m^2 \cdot s^{-2}$ indicate a possibility of supercells, with 300 supporting strong tornadoes and above 450 supporting violent tornadoes. A 0-1 km SRH of 100 is associated with moderate to strong tornadoes.

The depth of the SRH layer has an enormous influence on the resulting number, so it is important to select the correct depth. In many studies, the 0-3 km AGL layer is used. More recent studies (e.g. Craven and Brooks 2002) have found that the 0-1 km AGL layer has a greater correlation with tornado events. A 0-1 km bulk shear of 12 to 16 kt is a useful lower threshold for significant tornado events. It has been found, however, that locking a forecaster into fixed layers is not the optimal solution and in many cases may not accurately represent the streamwise vorticity being ingested by the storm. This has led to a quantity known as the effective SRH (Thompson et al 2005). This uses an effective inflow layer (q.v.) as the bounding layer for SRH computation. Storm movement may be computed using the ID method.

Though the area "swept out" by SRH on a hodograph is immediately obvious, an exact numerical value cannot be easily estimated. On the same token, it is not advisable to rely on simple numerical values of SRH without displaying the hodograph construction that led to this measurement. Such caution is required because SRH is sensitive to accurate storm motion estimation and accurate wind profiles. The ideal solution is to use a tool capable of showing both the detailed hodograph and an SRH value readout, such as AWIPS, Sharp, or RAOB.

Furthermore, the assumptions on storm motion, the exact SRH layer that is important to the storm, and local variations in the wind field, such as from mesolows, are all critical factors for an accurate SRH estimate. These values can change over time. Proper diagnosis of the atmosphere is essential for proper use of SRH.

10.12. SHEAR DEPTH. Some studies have explored shear depth: the layer where deep-layer shear is focused. Shallow depths are indicated by deep-layer shear which is focused in a shallow layer between the surface and the mid-troposphere, while large depths are indicated by deep-layer shear spread through the entire troposphere depth. Shallow shear depth appears to show some correlation

(Brooks et al. 1994) with strong, cold rear-flank downdrafts and is supported by numerical modelling.

10.13. STORM-RELATIVE MID-LEVEL FLOW. One indicator of supercell potential identifies moderate or strong environmental winds in the mid-levels of the troposphere *relative to the storm*. This can be estimated by starting at the tip of the storm motion vector on the hodograph and measuring the magnitude to an appropriate mid-level segment of the hodograph trace such as 5 km. One study found that a storm-relative mid-level wind exceeding 20 kt was associated with tornadic supercells.

10.14. STORM-RELATIVE ANVIL WINDS. This quantity indicates how well "ventilated" a storm is.

11. Instability-shear relationships

High instability and high shear are commonly associated with tornado events, however it has been found that these two elements frequently do not co-exist.

11.1. HIGH SHEAR - LOW INSTABILITY. When shear is very strong but instability is weak, one mode of storms that may form are mini-supercells. They are often observed in cool-season weather events as well as in locations on the periphery of tropical cyclones.

11.2. HIGH INSTABILITY - LOW SHEAR. High levels of pre-existing vertical shear are not necessarily a prerequisite for significant tornadoes. They sometimes occur in environments which are largely

What it takes for rotation

Recent work has settled on 0-1 km SRH and LCL height as two of the better discriminators between tornadic and nontornadic supercells, in environments that have sufficient deep-layer vertical shear and instability. Still, there are ways to improve on these simple ingredients when forecasting tornadoes. All SRH values are not created equal, and some evidence exists in support of the stronger shear being even closer to the ground (lowest 300-500 m) for significant tornadoes. The so-called "critical angle" measures the angle between the storm-relative inflow at the surface and the lowest layer shear vector - angles near 90 degrees mean the lowest level vorticity is almost completely streamwise, which results in the strongest rotation closest to the ground when this vorticity is tilted and stretched.

LCL height is related to the potential for relatively warm or cold RFDs, but the relationship is very complicated and depends as much (or more) on the types of precipitation particles in the hook echo region. Thus, LCL heights should be viewed in a probabilistic sense, where the chance for a tornado increases with lower LCL height (around 1000 m), and decreases with higher LCL height (1500-2000 m or greater). An environment with relatively high LCL heights (roughly 1300-1700 m AGL) can still produce a strong tornado if the other ingredients are clearly favorable.

Strong vertical shear through a deep layer of the atmosphere aids in the removal of precipitation from the updraft region, and promotes longer-lived storms. The oft-overlooked aspect of the deep-layer shear is the orientation of the shear vectors compared to the storm motion. "Classic" supercells have precipitation that falls out down wind of the mesocyclone and in the general direction of the storm motion. When this occurs, it sets up a situation where the storm inflow rides along the forward flank core and vorticity can increase due to small temperature differences between the core and the inflow. LP supercells tend to have limited rain cores that fall out well left of the storm motion. HP supercells usually have weaker flow aloft and much of the heavy rain falls near or behind the mesocyclone.

RICH THOMPSON, 2009
personal communication

devoid of shear but are very unstable. This poses the question of where the vorticity or shear comes from for tornado spin-up.

It is speculated that the storm either obtains vorticity from shear along a pre-existing boundary, which breaks up into a series of small-scale misocyclones with vertical vorticity, or produces its own horizontal vorticity roll which is later tilted by an updraft. In either case, a strong updraft is required which then stretches these vorticity sources and accelerates them. Intersecting boundaries can be an important source of vertical vorticity.

Low shear, because it provides a poor measure of updraft-downdraft separation, implies outflow-dominant modes, a process which is detrimental to tornadogenesis and results in non-steady storm states. It is speculated that low-shear tornadoes occur when an exceptionally strong updraft produces very strong inflow, temporarily suppressing or diverting the downdraft; *tornadoes may also be generated when instability is simply so extreme that any outflow-driven surface layer is inconsequential* (Davies-Jones et al 2001).

12. Moist symmetric instability (MSI)

The vast majority of forecasters are familiar with instability associated with very steep lapse rates, more properly, a profile in which environmental θ_e (equivalent potential temperature) decreases with height. This is called convective instability, also known as gravitational instability. The result is gravitational convection with vigorous, erect cumulonimbus clouds.

Moist symmetric instability is a different type of instability that can occur in stable air masses. It produces banded "slantwise" convection. *This is not an important process for severe weather* and some readers may wish to skip this subject, but since it can sometimes produce flooding rains, is frequently a source for weather along stationary fronts and other slow-moving boundaries, and describes precipitation processes in the stratiform region of the MCS, the topic may be of some interest.

Like gravitational instability, symmetric instability is not a process that actually powers convection. Rather it prepares an environment suitable for the development of weather. Moisture and lift are also required: adequate moisture, preferably a saturated layer, plus a source of synoptic-scale forcing to displace parcels in an up-

---

The difference: θ_e and θ_{es}

For a definition of theta-e (equivalent potential temperature), see any good thermodynamics textbook. Theta-es is the theta-e that air *would have* if it were saturated. The theta-es of unsaturated air can be measured and and it has meaning. If you look at the mathematical expression of theta-es in Schultz and Schumacher (1999, footnote 1), you'll see that none of the variables are functions of moisture. Theta-es is a function of temperature and pressure only. Theta-e, on the other hand, is a function of temperature, pressure, and moisture.

I believe this is where the confusion over "saturation" comes from. Schultz and Schumacher (1999) are advocating using theta-es to assess CSI because we are just following the traditional definitions for upright instability: when theta-e decreases with height, that is PI. When theta-es decreases with height (equivalently defined as the lapse rate being between the moist and dry adiabatic lapse rates), that is conditional instability. Therefore, *by definition*, CSI is assessed using theta-es and PSI is assessed using theta-e. There is no wiggle room to assert otherwise. Using theta-es does not imply that we make the assumption that RH is everywhere 100%.

DAVID SCHULTZ &
PHIL SCHUMACHER
CSI homepage, nssl.noaa.gov/csi

ward direction. The stronger the MSI, the smaller the wavelength and stronger the intensity of convective bands.

12.1. INERTIAL INSTABILITY. Inertially unstable areas occur wherever the absolute geostrophic vorticity is less than zero.

12.2. CONDITIONAL SYMMETRIC INSTABILITY (CSI). This refers to a specific type of thermodynamic pattern conducive to development of banded "slantwise convection". This type of convection occurs in marginally *stable* layers, is weak with little or no lightning, and rarely causes severe weather. Areas of CSI convection generally organize into mesoscale-beta scales of motion. The bands are oriented parallel to the thermal wind, thus roughly parallel to the surface front. If the lapse rates steepen and favor convective instability, then strong convection from convective instability will dominate instead of CSI.

Patterns favoring CSI are normally found in areas poleward of warm fronts where there is elevated ascent, especially where the layer is saturated and significant vertical shear is present. Soundings generally do not show instability. CSI patterns are enhanced by differential moisture advection and frontogenesis.

12.3. POTENTIAL SYMMETRIC INSTABILITY (PSI). Potential symmetric instability is a condition that exists when lift, such as from a synoptic-scale disturbance, will cause the atmosphere to develop conditional symmetric instability. To analyze for PSI, isopleths of θ_e (equivalent potential temperature) are overlaid against fields of M_g (geostrophic absolute momentum). Wherever θ_e decreases with height, the atmosphere is drying with height. This is conducive to destabilization because if the entire column is forced to rise, latent heat will be released more rapidly in the lower levels than in the upper levels, which will increase the lapse rate.

12.4. CSI ANALYSIS. To analyze for CSI, isopleths of θ_{es} (saturated equivalent potential temperature, sometimes written as θ_e^* with asterisk) are overlaid against fields of Mg (geostrophic absolute momentum, also geostrophic pseudoangular momentum) on a cross section that is roughly perpendicular to the temperature gradient.

Normally the momentum isopleths tend to be oriented more vertically than the θ_{es} isopleths, but where the momentum isopleths are less vertical this indicates areas of CSI. After locating CSI, the CSI cross-section is evaluated against the cross-section of relative humidity field to see if the instability will actually result in weather. The presence of synoptic-scale lift or subsidence must also be considered.

Values of θ_{es} are strictly a function of temperature and pressure. Wherever θ_{es} decreases with height, the lapse rate in that layer is conditionally or absolutely unstable. Where θ_{es} increases with height, the lapse rate is stable.

The simplest expression of absolute momentum is $M_g = v_g - f$, showing that it equals the geostrophic wind speed minus Coriolis effect. The term v_g is normally stronger aloft, and f is higher in the polar regions, so M_g is normally highest aloft and in equatorward latitudes. A parcel tends to be inertially stable and conserve its momentum. As seen on a cross section, parcels tend to "cling" to isopleths of M_g, accelerating back to them if displaced.

Therefore to get large amounts of CSI, it is necessary for the M_g isopleths to be oriented horizontally rather than vertically. The best way to do that is have a strong increase in wind speed with height (speed shear) so that momentum increases dramatically with height. With this configuration, their slope more easily exceeds that of the θ_{es} isopleths.

It is also possible to evaluate CSI, which occurs in areas of near-zero equivalent potential vorticity (EPV), but a discussion of this is beyond the scope of this book. When the EPV is below zero, then PSI or convective instability is present.

12.5. CSI STRENGTH. In the same way that CAPE determines the energy in gravitational convection, SCAPE (slantwise convective availability of potential energy) allows the energy in slantwise convection to be determined. Some preliminary studies, however, have found questionable results from SCAPE measurements. Some research suggests it is better to simply solve for $\text{MPV}_{gs} = M_g - \theta_{es}$ which yield the lapse rate available to the slantwise parcel. The more negative the MPV_{gs} the stronger the slantwise convection.

12.6. LAYERS WITH CSI QUALITIES. In saturated layers where M_g and θ_{es} contours are parallel, the layer is said to have weak symmetric stability (WSS). The type of convection or weather that takes place here is indeterminate. Slantwise convection may occur in a WSS layer if frontogenetic forcing is available.

Any layer in which θ_{es} actually decreases with height indicates the presence of conditional instability, CI. With saturation in this layer, it will result in the typical deep, buoyant convection described elsewhere throughout this book. Even though momentum isopleths may in fact be less sloped than θ_{es} lines in this area, the slow development of CSI bands are largely destroyed by the relatively strong vertical motions associated with buoyant convection. Areas of convective instability are common on the equatorward side of CSI regions.

12.7. CSI ANALYSIS CAVEATS. CSI analysis is hindered by the synoptic-scale nature of radiosonde observations, which in terms of size and duration does not lend itself to the mesoscale nature of slantwise convection. Also since it is dependent on expressions of the geostrophic wind speed, results will not be valid in ageostrophic flow, including in areas of cyclonic or anticyclonic curvature. Helicity in the layer, indicated by a shear vector that turns with height, also invalidates some of the assumptions in the CSI analysis.

13. Composite parameters

To grasp the highly complex nature of storm environments, many forecasters use simplified, easily-understood representations of temperature, moisture, and shear. These representations are known as diagnostic variables. They can be extremely useful for providing a quick summary of one particular aspect of the storm environment.

An index is perfectly accurate at measuring the specific quantity it represents, such as the temperature difference between two levels. However the correlation of the parameter to severe weather is never perfect, and with some indexes its skill is poorly researched or is largely anecdotal. Inappropriate use of a diagnostic variable is poor science and can lure the forecaster into discarding key parts of the forecast process.

Diagnostic variables in common use today are reviewed here for their forecast use. Detailed definitions and other indices are listed in the appendix. All of them require proper modification of the sounding. ¤

8

RADAR

The Army Air Forces announced today that radar had been used as a foolproof weather prophet, forecasting hurricanes hours before their arrival. Lt. Col. Harry Wexler of the AAF Weather Service told a meeting of the American Meteorological Society and the Institute of Aeronautical Science at Columbia University that last September's Florida hurricane was [tracked] by radar 20 hours before it struck. He explained that discovery of radar's weather forecasting abilities was discovered almost by accident after pilots frequently "flew themselves dizzy" searching for enemy aircraft only to discover their radar screen was recording weather changes.

— International News Service
January 31, 1946

Weather radar offers an unprecedented way to look at precipitation. This technology uses a rotating antenna to emit an extremely brief but intense pulse of radio energy, then listens for the reflection of this energy. The direction that the dish is pointing yields azimuth and elevation, while the time it takes for radar energy to return to the dish yields range, a technical term for distance.

Up until the 1990s, most radars swept through 360 degrees with the antenna at a fixed elevation, usually 0.5 deg above the ground. The radar would operate this way all the time except when the radar operator was scrutinizing an echo. Nowadays, most network weather radars are computer-controlled and scan several different antenna elevations in order to capture a three dimensional look at the atmosphere. Each complete set of scans at multiple elevations is called a *volume scan*. With the WSR-88D one of these volume scans is completed every 5 to 10 minutes.

1. Base products

The base products of a conventional radar include reflectivity, velocity, and spectrum width. Only these three fields are needed to produce displays and a plethora of other derived products.

1.1. BASE REFLECTIVITY (R). Reflectivity is the most familiar radar product. It is a measure of the power returned, or backscattered, to the radar from solid and liquid targets. It must be remembered that the radar is far more sensitive to precipitation than to clouds, which are composed of much smaller drops and are barely detectable by conventional radars. As a result, radar will be quite effective at detecting precipitation-filled downdrafts, but updrafts, whether in the form of clear thermals or cumuliform towers, will generally not show up on radar until precipitation is produced.

1.2. BASE VELOCITY (V). By measuring the slight shift in radio frequency of backscattered radiation, it is possible to estimate the velocity of any given target towards or from the radar. The principle is similar to the change in pitch of a car as it passes an observer: the approaching car makes the sound waves pass at a fast rate, and as

Human radar
Germany can be credited for establishing a sort of radar network before the technology was even invented. During the 1909 Aeronautical Exposition in Germany, the government created a network of 80 human observers in the region surrounding Frankfurt-am-Main. These observers watched for thunderstorms and sent detailed telegrams indicating time, bearing, and movement. The central office used triangulation to map each storm in real time and predict its course and speed. Storms that were bearing down on the expo prompted a warning for event coordinators. It was such a success that by 1912 the country established a real-time warning network using hundreds of government postmasters.

Figure 8-1. WSR-88D network radar at Oklahoma City, Oklahoma (TLX), September 2009. *(Tim Vasquez)*

it departs the sound waves pass at a slow rate. This is known as the Doppler effect.

The WSR-88D (NEXRAD) and most modern Doppler radars detect this change in frequency, but do not directly calculate the frequency shift. The speed of a downdraft is about a trillionth of the speed of light, too small of a fraction to accurately measure. However it is possible to transmit two separate radar pulses and actually look at the incoming radar energy as a waveform measuring hundreds of miles in length. Using this wave, we can look for shifts in the phase of this wave caused by motion in various weather targets. If the weather target is stationary, the phase of both reflections coming from the target will be the same; in other words, the crest and trough of the reflected radio energy will be exactly identical compared to the transmitted radio energy. However if the target is moving, the crests and troughs will differ, indicating a change in the phase of the waveform.

The technique of comparing these two waveforms is called *pulse-pair processing*. It is used in the WSR-88D radar, and while it provides a novel way to obtain velocity, it imposes some limitations on measurements, such as aliasing, all of which will be discussed shortly.

1.3. SPECTRUM WIDTH (SW). Spectrum width measures the diversity of velocity within a specific radar bin. If all the scatterers are moving the same direction and speed, the spectrum width will be low. Scatters moving chaotically in the bin, on the other hand, will produce high spectrum width values. This product has long been neglected in operational meteorology, and many end-user web sites specializing in radar do not even offer it. Some attempts, however, have been made to identify small tornadoes with this product. Some recent research has demonstrated that three-body scatter spikes from hailstorms may be more visible on spectrum width than on reflectivity products. The most recent research has indicated that it may be useful for finding updraft cores, which in turn have a bearing on hail production characteristics.

2. Limitations of radar

Though radar provides an amazing tool for severe weather forecasters, basic physics and technological considerations impose some problems on the use and interpretation of radar. All forecasters should be familiar with these issues.

2.1. EXCESSIVE HEIGHT. The higher the antenna elevation or the more distant the storm, the higher the radar bins will be within the storm. The effect on distant storms is a consequence of the earth's curvature. If elevated slices of a storm are used, precipitation fields and important circulations close to the ground will not be seen. For this reason most forecasters favor the lowest possible elevation (0.5 deg on the WSR-88D; anything lower than 0.5 deg causes the conical beam to intersect the ground). Higher elevations are used to view structures within the storm. Forecasters should always double-check the radar elevation before interpreting a product to be aware of which height in the storm is being examined. It should also be noted that the WSR-88D product known as composite reflectivity is made up of backscatter at all heights at a given point, and inattentive use of it may cause elevated echoes to mask important lower-level echoes.

2.2. EXCESSIVE RANGE. As it is impossible for the radar beam to match the Earth's curvature, the radar is unable to see the lower portions of distant storms and in many cases may overshoot these storms. As a result, important weather phenomena may be missed. Furthermore the spreading of the radar beam makes fine detail unresolvable. With the WSR-88D, this spreading widens to about 2 nm at 120 nm. The useful limit of the radar for tornado detection is about 60 to 80 nm as the radar bins at this range are too wide and too high to pick up tornado circulations. Most storms are undetectable past 250 nm since the lowest possible elevation of the radar beam will overshoot all but the tallest storms.

It's radar time
In a typical weather pattern, a WSR-88D is programmed to spend 1.57 microseconds emitting a pulse of energy and then an incredibly long 3145 microseconds listening for the reflection of that pulse. One microsecond is a millionth of a second. Since light travels 161875 nm each second, simple division tells us that the pulse has a length of 0.25 nm and travels 509 nm between each pulse. Thus it can make a round trip to a cloud 254 nm away, a total distance of 509 nm, before the next pulse is emitted.

A glimpse into the future

It is anticipated that radar will provide useful information concerning the structure and behavior of that portion of the atmosphere which is not covered by either micro- or synoptic-meteorological studies. We have already observed with radar that precipitation formulations which are undoubtedly of significance occur on a scale too gross to be observed from a single station, yet too small to appear even on sectional synoptic charts. Phenomena of this size might well be designated as mesometeorological.

MYRON G. H. LIGDA, 1951
first use of a 'mesoscale' phrase

2.3. EXCESSIVE PROXIMITY. Proximity to the radar is important for measuring mesocyclone and tornadic circulations. But there are also serious consequences when a storm is too close to the radar. As the WSR-88D antenna does not tilt higher than 20 degrees, scatterers within 20 or 30 miles of the radar and in the mid- or upper-levels of the troposphere may go completely undetected. Echo tops and VIL are the first to suffer from close range since they sample only the bottom parts of a typical storm. Hail algorithms and storm tracking are also degraded by this effect.

2.4. RANGE FOLDING. The radar must transmit pulses frequently enough so that the entire volume scan can be completed accurately in a short amount of time. But if radar information from one pulse arrives at a distant storm, reflects back to the radar, and arrives after the next radar pulse is emitted, it will be erroneously processed as part of the second pulse. As a result, the echo will be plotted in a false location close to the radar. This will also degrade real echoes which might be arriving at the same time from the second pulse. To avoid this, the radar can use a scan strategy with low pulse repetition frequency, allowing the pulses to travel farther before sending another one, however this degrades the radar's velocity products as we shall discuss shortly.

2.5. PRECIPITATION EFFICIENCY. Low-precipitation (LP) supercells have a weaker appearance on radar owing to the lack of precipitation particles, and may have no concavity or hook echo. There may be no signature at all to identify the updraft area. Classic (CL) supercells, however, have all of the textbook features of a supercell, including hook, concavities, WER/BWER, and observable mesocyclones. At the far end of the spectrum, high-precipitation (HP) supercells contain extensive radar-observed precipitation areas. The hook is very well-defined and may be quite large, giving the storm more of a C-shaped appearance.

2.6. VELOCITY ALIASING. The WSR-88D's pulse pair processing technology makes it susceptible to an effect called *aliasing*. This is effectively a speed limit, called the Nyquist velocity, which is imposed on the interpretation of the waveform (see the special feature

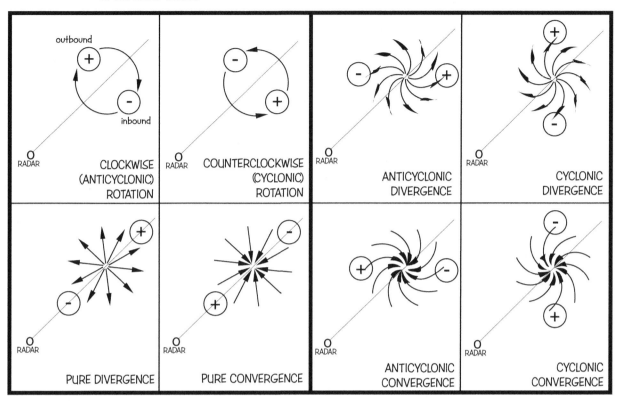

page here on aliasing). In short, the range of detectable velocities corresponds to a single sine wave occupying the range -180 to +180 degrees, with 0 indicating a velocity of zero and -180 and +180 indicating the incoming and outgoing Nyquist velocity. If a parcel exceeds the Nyquist velocity dictated by the scan strategy in use, the target's phase shift exceeds the limits of -180 to +180 degrees and is mapped to the next full wave, i.e. a shift of 192 degrees will be treated as -168 degrees. As a result, the processed velocity will "fold" over to the other side of the velocity spectrum.

In summary, the WSR-88D and other pulse-pair processing systems have boundaries on the maximum detectable winds. The radar can increase the Nyquist velocity by boosting the pulse repetition frequency, but this will limit the range of the radar and increase range folding problems. The WSR-88D now employs a dual scan strategy where it uses low PRFs for reflectivity and high PRFs for velocity. This allows for some improvement, but aliasing and range folding are still seen frequently.

Figure 8-2. Reference chart for determining motion type by observing the orientation of a velocity couplet relative to a radar site. Velocities which are positive (+) are outbound, i.e. moving away from the radar, while velocities which are negative (-) are inbound, i.e. moving toward the radar. Learning these signatures is an essential to Doppler radar forecasting.

A short tutorial on aliasing

Imagine that instead of a Doppler radar you have a cabin out on the prairie and a telescope aimed at a large, fixed wind turbine up on a ridge. You've noticed that by counting revolutions, you can guess the wind speed. One arm of the turbine blade is black and the other white, making it easy to count the turns without getting confused. You've found that when a blade makes a full 360-degree turn in one second, this means the wind is blowing at 100 mph. When the blade spins clockwise winds are blowing toward us (inbound), while counterclockwise spin means winds are blowing away (outbound).

Now it is nighttime and a storm is brewing. But you can only illuminate the anemometer with a giant Army surplus xenon strobe light. When you push a red button you get two brilliant flashes, exactly one second apart.

All ready? Let's take measurements. We press the big red button. With the first flash the turbine blade is at the "12 o'clock" position, and with the second flash we see the blade at the "3 o'clock" position. That's a 90 degrees clockwise change. What does this mean? To complete a full 360-deg turn it would take the blade 4 seconds, so this implies an inbound wind speed of 25 mph. So far, so good.

The storm strengthens so we wait a minute, recharge the strobe light, and press the red button again. With the first flash the blade is at the top, and with the second flash the blade is at the bottom. That's a spin of 180 degrees, or 50 mph. But wait a second. Which direction? Was it 180 degrees counterclockwise or 180 degrees clockwise? Based on our observation alone, we don't know! The wind is certainly blowing 50 mph but could be either inbound or outbound.

The storm continues and the lightning increases. We take another measurement and find the blade moves from "12 o'clock" to "9 o'clock", a 270 degree clockwise change. That's 75 mph! But a long stroke of lightning shows a tree on the ridge waving gently, leaning away from us. We were fooled! After thinking about this we realize the turbine actually rotated counterclockwise by 90 degrees, thus a 25 mph outbound wind occurred.

What we've learned in this experiment is that winds strong enough to turn the turbine exactly 180 degrees in our sampling interval will lead to ambiguity in wind measurements. A 180 deg turn in 1 second equals a 360 deg turn in 2 seconds, thus 50 mph winds. So the Nyquist velocity, or maximum unambiguous velocity, of our apparatus is 50 mph. The range of wind speeds we can sample, or the Nyquist co-interval, is 50 kt inbound to 50 kt (-50 to +50 kt). When the winds exceed this velocity they will produce ambiguous results that will cause confusion.

A better idea is to emit faster strobe flashes, in other words, change the pulse repetition frequency or PRF, so we can catch the spin changes before they rotate through 180 deg. We calibrate our strobe to pulse 0.5 seconds apart. Now if the winds are strong enough to rotate the turbine 180 degrees during this new 0.5 second interval, that implies a speed of 360 deg per second, or 100 mph. So we have boosted our Nyquist velocity to 100 mph and can accurately sample storms with a range of -100 to +100 mph.

For a Doppler radar, the Nyquist velocity is determined by the basic equation $V = P*W / 4$, where V is maximum unambiguous velocity (m s^{-1}), P is pulse repetition frequency (Hz), W is wavelength (m). So for the WSR-88D, which has a wavelength of 0.107 m (10.7 cm), a typical PRF of 1304 Hz, this yields a maximum unambiguous velocity of 34.882 m s^{-1} (67.8 kt). The lowest PRF is 318 Hz, which yields a maximum unambiguous velocity of 8.5 m s^{-1} (16.5 kt).

While high PRF is always better for Doppler measurements since it increases the Nyquist velocity, it causes the radar pulse to cover a shorter distance before another pulse is emitted. Since light travels 161875 nm per second, a PRF of 1304 Hz only allows the pulse to travel 124 nm before another pulse is transmitted. Any backscatter that comes from targets outside this radius get mapped to the wrong range, producing range folding artifacts. Boosting the PRF to the WSR-88D's maximum of 318 Hz increases the effective range of a pulse to 509 nm but lowers the Nyquist velocity. This tradeoff is commonly called the "doppler dilemma". While the dilemma is an unalterable effect of the laws of physics, there are ways to work around it, such as using scans with variable PRF, using dealiasing algorithms, and using range unfolding algorithms.

Figure 8-3. Aliased velocities in a tornadic storm. The tornado location is marked with a circle. The radar was operating in VCP 11, which in this case had a maximum unambiguous velocity of 52 kt. Storm motion was quite close to this limit. This storm contains a core of aliased velocities (dark color on left), which shows as strong outbound embedded within a field of strong inbound. An

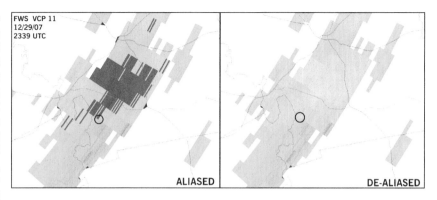

untrained forecaster looking at the uncorrected display (left) would identify a false boundary of strong anticyclonic shear on the storm's south side. De-aliasing on the GRLevel2 program (right) more accurately reveals the actual velocity fields, barely showing the cyclonic couplet at this distant range.

Figure 8-4. Derecho storm with raw base velocity (lower left) and storm relative velocity (lower right) with dealiasing. The leading edge of the derecho is marked with the thick dashed line. In the vicinity of the derecho, very high inbound velocities on the order of -50 kt are displayed, which indicates strong outflow, and is painted in a bright green color. However this immediately gives way to the red outbound area behind the derecho, marking an immediate transition to +50 kt outbound. This immediately signals to the forecaster that the entire outbound area is aliased. With dealiasing and compensation for storm motion (lower right) the outbound velocity area has been correctly changed into a very strong inbound velocity area, but the dealiasing has not been entirely successful as the tail end of the derecho is poorly handled. Since dealiasing functions in various software programs do not always do a reliable job, inspecting the base velocity productions for aliasing and range folding issues can help the forecaster be aware of problems in derived products. The ring on the left is range folding at the 74 nm maximum unambiguous range for the VCP 212 scan strategy being used.

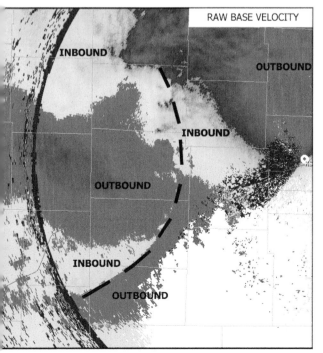
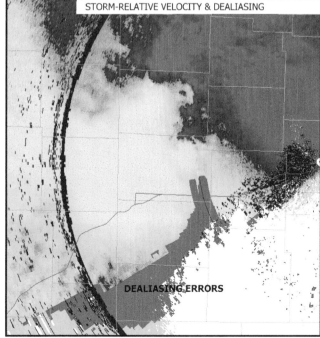

2.7. REFRACTION. In the same way that a laser beam will change direction when it enters the water in a swimming pool, a radar beam is affected by unusual changes in density (temperature) with height. The most common effect is superrefraction, which occurs when a strong inversion is present. This causes the beam to bend more strongly toward the ground and may cause interference from ground clutter. On the other hand, when the lapse rate is very strong, as in a very unstable atmosphere, subrefraction may occur with the beam bending more strongly upward than usual. This causes echoes to come from higher parts of the storm than expected and may result in distant storms being missed.

2.8. BEAM BLOCKAGE. If the beam is blocked by mountains or nearby buildings, it may result in radial gaps in coverage. Forecasters should be familiar with each radar commonly used and be aware of areas that are often blocked. NOAA publishes graphics at <http://www.ncdc.noaa.gov/nexradinv/> under "coverage maps" which show where terrain blocking occurs at each site.

2.9. WAVELENGTH. The WSR-88D is a 10 cm long wavelength radar. This is in the S-band according to IEEE convention and is considered the standard frequency band for long-range weather surveillance. The TDWR and many TV station radars are a 5 cm (C-band) radar. Many onboard boat and aircraft radars are 3 cm (X-band) short wavelength radars. These are very important differences to consider when using an unfamiliar radar source. Long wavelengths are more sensitive to larger particles, are not easily attenuated by precipita-

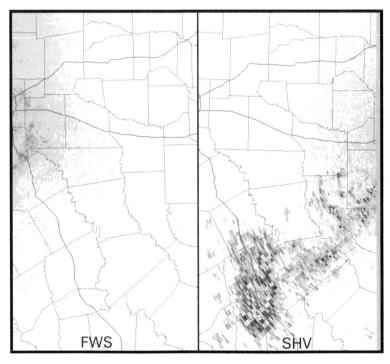

Figure 8-5. A large cluster of showers in east Texas? Interrogation of an apparent shower area (right) by another radar in the area (left) shows that it's just anomalous propagation. This can be confirmed by surface data, satellite imagery, and a sound diagnosis of the weather situation. *(Tim Vasquez / graphics GRLevel2 <grlevelx.com>)*

tion, and are not very susceptible to velocity aliasing errors, but are not sensitive to small cloud droplets and require larger, expensive equipment with greater power consumption. Short wavelength radars are more sensitive to small cloud droplets such as drizzle and light rain and are cheaper and smaller. However they are attenuated more easily, and with Doppler versions of these units the short wavelength increases the problems with velocity aliasing.

3. Polarimetric radar

The next major advance of weather radars, expected to form the backbone of weather radar networks worldwide in the 2010s and 2020s, will measure the polarization characteristics of back-scattered radiation. Just as driving on a bumpy road produces vertical movements that force the driver to bounce up and down in his seat, the bumpy road can also have side-to-side oscillations. If the driver is blindfolded, there is much to be learned from all of these oscillations!

While conventional radars only measure electromagnetic oscillations in one plane, a *dual-polarization radar* measures these quantities in both horizontal and vertical planes and can compare and contrast the two. This yields information on particle size, shape, orientation, and dielectric strength. This class of radar is also known as a *polarization diversity radar*. When energy is emitted from a radar antenna in a particular plane, it typically reflects off precipitation in the same plane. This energy when measured is referred to as co-polar power. However an extremely small amount of energy will be reflected in the perpendicular, or orthogonal plane. When measured, this is referred to as cross-polar power.

Considerable groundwork exists for this technology. The first dual polarization radars were developed in the 1970s the NEXRAD tri-agencies began experimenting with a WSR-88D upgrade in 2002. The network was about to be modernized at this book's publication time effective with NEXRAD Build 12, and polarimetric data should be widely available to forecasters, hobbyists, and researchers in the United States starting in 2010.

The main benefits of the technology are accurate discrimination of precipitation types, better rainfall measurement, and distinct identification of hail. Dual-polarization radar proves better precipitation estimates, eliminates problems with bright band pat-

---

The true value of radar

An elaborate map was prepared by the editor in 1897 as a preliminary step toward the collection of thunderstorm data, and the organization of a system of daily thunderstorm predictions for [Washington D.C.] Every telegraph and telephone station within a hundred miles north, south, and west, was plotted down, and it was quickly found that thunderstorms whose average diameter is five miles would, inevitably, slip through when approaching from the northwest, and could rarely by detected when approaching from the west, or the north, the southwest or the south, in time to allow any satisfactory prediction.

Stations must be within a mile of each other in all directions in order to catch every tornado and determine the direction of its path in time to frame a warning that could be of any use to a central city. We have no right to issue numerous erroneous alarms. The stoppage of business and the unnecessary fright would in its summation during a year be worse than the storms themselves, so few and so small are they . . .

CLEVELAND ABBE, 1899
editor, Monthly Weather Review

Figure 8-6. Bounded weak echo region (BWER) in the May 5, 1995 "Mayfest" HP storm about half an hour before it swept through Fort Worth. The views are a standard 0.5 degree elevation (right) and a cross section along a west-east line (below). The 0.5 degree elevation at right corresponds to the bottom of the cross section. The BWER is the upside-down U-shaped concavity at about 15 to 30 thousand feet MSL, positioned above the updraft area. Note that the 0.5 degree slice at right gives no hint as to how massive and ominous the overhang and BWER is. It is also noteworthy that the highest base reflectivity, which is fixated on by most forecasters, is not underneath the bulk of the intense echoes. *(Tim Vasquez / graphics GRLevel2 Analyst <grlevelx.com>)*

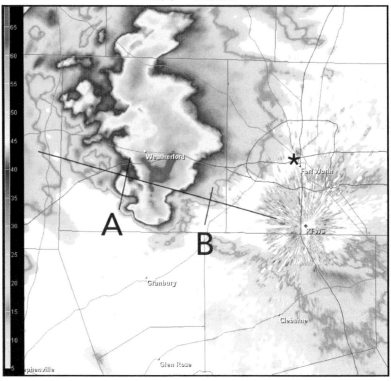

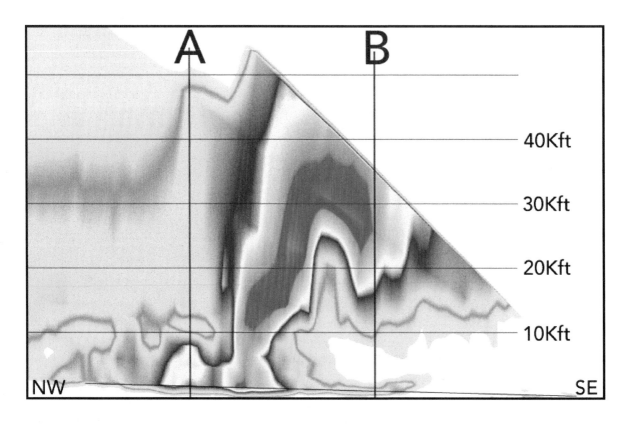

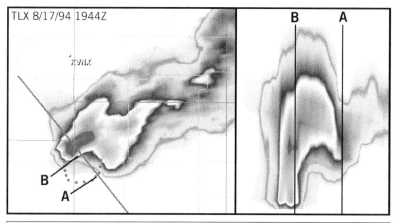
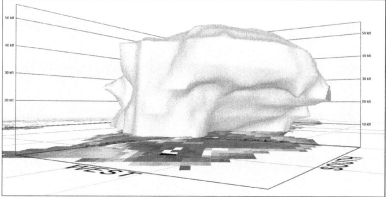

Figure 8-7. The 1994 Lahoma storm produced 113 mph winds at an Oklahoma mesonet station on 17 August 1994. Viewed on radar, a cross section seems to show a bounded weak echo region. However other cross sections reveal that the vault is not enclosed. A 3-D depiction of the same image (below) from GrLevel2 Analyst reveals that this structure is simply a WER, or overhang. This demonstrates the value of attention to detail when visualizing radar structure. *(Tim Vasquez / graphics GrLevel2 Analyst <grlevelx.com>)*

terns, eliminates anomalous echoes due to birds and insects, and mitigates terrain blockage problems. A study during JPOLE compared polarimetric hail detection to existing WSR-88D methods and found that the probability of detection increased from 88% to 100% and the false alarm ratio dropped from 39% to 11%.

4. Polarimetric base products

Conventional radars offer three base products: reflectivity, spectrum width, and velocity. Dual-polarization radars offers all three plus four additional products: differential reflectivity, correlation coefficient, differential phase shift, and specific differential phase. These are described as follows:

Figure 8-8. The radar analyst.
(David Hoadley)

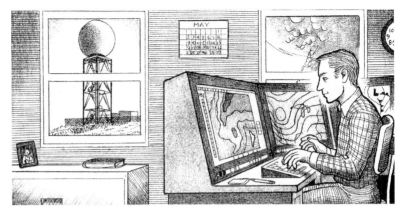

4.1. BASE REFLECTIVITY (Z_H, Z_{HH}), dB. The display of reflectivity from the radar polarized on the horizontal plane (Z_H), provides exactly the same information as on non-polarimetric radars.

4.2. BASE VELOCITY (V), kt. Base velocity data on polarimetric radars is the same as that on non-polarimetric radars.

4.3. SPECTRUM WIDTH (SW), dB. Spectrum width on polarimetric radars is the same as that on non-polarimetric radars.

4.4. DIFFERENTIAL REFLECTIVITY (Z_{DR}), dB. This measurement determines the ratio between power reflected in the horizontal plane from power reflected in the vertical plane. In short, $Z_{DR} = 10 \log (Z_h / Z_v)$. This formula shows that when horizontally-polarized reflectivity is dominant, ZDR is positive, and when vertically-polarized reflectivity is dominant, ZDR is negative. Typical values are -2 to +6 dB. This quantity gives information on the shape of the drop. For example, large droplets tend to be oblate and fall with a horizontal orientation (Z_{DR} of +2 to +4), while snow and ice tends to be close to 0 and hail tends to be spherical or oblate with a vertical orientation (i.e. Z_{DR} of 0 to -1). Conical ice shapes favor negative Z_{DR} values. Conversely, insects will have very high Z_{DR}. Debris from tornadoes tends to have a Z_{DR} close to zero. A drop in Z_{DR} in a winter storm pattern can indicate a transition from rain to snow.

4.5. CORRELATION COEFFICIENT (ρ_{hv}) (rho$_{DP}$) (CC), ratio. This measure examines all of the pulses within a radar bin to determine the distribution of power between horizontal and vertical planes. High ρ_{hv} indicates low diversity, while low ρ_{hv} indicates high diversity. Values of 0.96 to 1 are common in rain and dry snow. Values of 0.85 to 0.95 indicate a mix of different precipitation types, including hail and aggregates. Below 0.85 other phenomena are likely taking place, such as ground clutter, insects, birds, chaff, and at even lower values, debris.

4.6. DIFFERENTIAL PHASE SHIFT (ϕ_{DP}) (phi$_{DP}$), deg. Measures the difference in phase between the horizontal and vertical plane at a particular radar bin. Radar energy may slow very slightly as it passes through water and ice, since the speed of light in those substances is different from that in air. While this slowing is not readily detectable, it can be immediately apparent when the waveforms of the horizontal and vertical phases are compared. If the waves are out of sync, that is, out of phase, this indicates that one of the phases was slowed by water, air, dust, or other particles more than the other phase.

The phase difference is additive with range, with the phase shift at a given bin prone to greater and greater variance with distance. This can be likened to building a wall by stacking bricks vertically, one atop the other without mortar; since the top of the wall is entirely dependent on the thickness of bricks underneath, the bricks get more and more uneven with increasing height. This same quality gives the differential phase shift display a radially streaked appearance with greater "noise" the further one goes from the radar. It makes the product difficult to use, so forecasters generally do not use differential phase shift product, relying instead on a different base product called specific differential phase, listed below.

4.7. SPECIFIC DIFFERENTIAL PHASE (K_{DP}), deg per km. This quantity examines the differential phase shift (see above) and simply shows the amount of change in phase shift at each bin along the radial, rather than the total shift. This is expressed as phase change per unit distance. As a result, the specific differential phase product helps highlight where phase shifts are occurring and where they are most intense. Positive K_{DP} indicates that energy has been slowed

WSR-88D HDA

- Probability of Hail (POH). The POH quantity estimates the risk of any size of hail reaching the earth's surface. It primarily looks for the height of the 45 dBZ echo above the melting level, raising the risk as this height differential grows.

- Probability of Severe Hail (POSH). The POSH algorithm estimates the amount of reflectivity above 45 dBZ that exists above the -20°C level. POSH works best in a sheared cool-season air mass, and will overestimate the risk in a weakly-sheared pulse storm environment.

- Maximum Expected Hail Size (MEHS). Finally the MEHS uses only the SHI data and simply increases the MEHS as the SHI grows.

- Performance and reliability. Overall, the HDA is well-regarded and one study done at NWS Wichita shows that it overestimates hail size but works very well with storms that develop in strong mid-level flow. The HDA is more likely to fail in processes that distort or obscure radar-observed storm structures: most notably high shear that causes storm structure to depart from the vertical, squall lines, which make individual storm structures difficult to resolve, and storms close to the radar site, whose upper portions are not sampled at all.

Dual-pole hydrometeors

A handy reference to precipitation categories:

GC/AP. Ground clutter/anomalous propagation.
BS. Biological scatterers.
DS. Dry aggregated snow.
WS. Wet snow.
CR. Crystals.
GR. Graupel.
BD. Big drops.
RA. Rain (light and moderate).
HR. Heavy rain.
HA. Hail, possibly mixed with rain.
Unknown.

mostly in the horizontal plane due to hydrometeors dominated by horizontally oblate particles like rain. Negative K_{DP} indicates slowing of energy in the vertically plane, and suggests vertically oblate particles like conical ice.

It has been found that K_{DP} is a very effective indicator for rainfall rate when very high basic reflectivity values are correlated to high K_{DP}, indicating significant backscatter from particles that are almost entirely raindrops.

5. Polarimetric radar derived products

Derived products are not directly detected but rather are based upon one of the seven base polarimetric products. Some of the products most likely to enter widespread use are as follows:

5.1. HYDROMETEOR CLASSIFICATION ALGORITHM (HCA). This algorithm attempts to identify the dominant type of precipitation at each radar bin. It uses all of the dual-polarization base products to make the determination. From that it selects from 11 different categories.

5.2. MELTING LAYER DETECTION ALGORITHM (MLDA). This algorithm examines dual-polarization volume characteristics to dry to determine the height of the melting layer. On meteorological systems such as AWIPS it is expected to be displayable as a series of 4 concentric rings on any horizontal product, with the inner two rings indicating the bottom of the melting layer and the outer two rings indicating the top.

5.3. QUANTITATIVE PRECIPITATION ESTIMATE (QPE). The QPE, comparable to the storm precipitation algorithm on the legacy NEXRAD unit, is expected to be highly accurate since dual-polarization algorithms can determine and factor in precipitation type. From a severe storm forecasting viewpoint, this will yield improved precipitation totals for flash flood forecasting.

6. Dual doppler radar

Dual doppler radar, not to be confused with dual polarization radar, refers to the use of two Doppler radars which attempt to observe a storm at oblique or orthogonal angles. This works around the inability of a single radar to observe velocity perpendicular to the beam. In the United States, this method is expected to remain limited to research programs for some time and will not likely be used by most operational forecasters for the foreseeable future.

Accuracy of dual doppler observations are highly sensitive to the location of the storm around the two radars. If the storm is parallel to or perpendicular to the radar baseline, then the radars observe the storm at similar angles and it becomes impossible to determine perpendicular aspects of velocity. The problem can be overcome by adding more radars to the observation domain to allow storms anywhere in the domain to be observed at perpendicular angles, but this adds geometrically to the cost of the network.

7. Storm evolution principles

The initial thermal and fair weather stage of the cumulus cloud is generally not detectable by conventional radars. As a towering cumulus cloud grows and large cloud droplets begin forming, however, radar reflectivities begin appearing at a height of about 3 to 5 km and increase to 0 to 20 dBZ. In many cases this is above the lowest radar elevation, so higher elevations or, better yet, composite reflectivity (to be discussed shortly) can be of great help in picking out these early signs of deep convection.

As the cumulonimbus begins maturing, the most intense core is found at the -10°C to -20°C layer (roughly around 20,000 ft) where supercooled drops, graupel, ice, and even hail are present. In showers and weak storms this core is usually not be much stronger than 30 or 35 dBZ, but in strong storms reflectivities of 50 dBZ and higher are common. As the storm continues evolving and the downdraft becomes well-developed, this intense core shifts downward to the ground and becomes stronger on the lowest radar elevation slices.

Warm cloud storms, that is, with no part colder than 0°C, often occur in the tropics and subtropics. Precipitation production is dominated by collision and coalescence processes. These are much

---

Radar classification

Gallus et al 2008 proposed a scheme for radar classification of echoes. These types are not necessarily different in terms of structure and weather, but the scheme is useful for categorizing what's seen on radar.

■ **Isolated cells (IC)**. These cells develop at some distance from other storms and are not along a common line.

■ **Cell clusters (CC)**. Cell clusters are storms whose strong areas are not connected, and the area does not show linear organization. Common when capping, instability, and flow are weak.

■ **Broken line (BL)**. The broken line configuration is associated with many supercell outbreaks. It tends to join up into a squall line over time.

■ **Squall line, no stratiform rain area (NS)**. Signifies an MCS developing in a dry environment.

■ **Squall line, trailing stratiform area (TS)**. This is the classic textbook MCS storm and is quite common.

■ **Squall line, parallel stratiform area (PS)**. May include embedded lines.

■ **Squall line, leading stratiform (LS)**. May occur in high-shear environments.

■ **Bow echo (BE)**. Occurs in an MCC which acquires a strong rear inflow jet.

■ **Nonlinear system (NS)**. In a non-linear system, the strongest echoes are connected but do not show linear characteristics. This includes clusters of embedded storms.

> **Embedded bow echoes**
>
> Because MCS systems are frequently associated with extensive areas of precipitation, a bow echo may actually be embedded in lower reflectivities and hard to see. Don't just fall for looking for the telltale yellow arc of higher reflectivities.

more efficient than methods involving ice. As a result, rainfall rate techniques and rules of thumb based on reflectivity will often underestimate precipitation.

8. Severe storm signatures

When discussing severe storm signatures in general, this addresses factors that indicate an intense strong updraft or downdraft, rather than specific phenomena like hail, wind, and rain. Some of the key signatures are described as follows.

8.1. INTENSE REFLECTIVITY. One indicator of a severe storm is very high reflectivity, which is often associated with severe storms. The traditional interpretation of high reflectivity is that tremendous amounts of precipitation and ice are present, produced by a significant updraft. However many weak storms have *efficient* precipitation characteristics and may readily produce large amounts of graupel and snow within the cloud, while many severe storms can be *inefficient* rain producers and generate little ice.

8.2. REFLECTIVITY GRADIENT. Strong reflectivity gradients are indicative of severe weather. A reflectivity core which is well-centered within the storm reflectivity mass will have a concentric appearance and exhibit weak reflectivity gradients. A core that has shifted to one side of the reflectivity mass, however, will produce a strong reflectivity gradient.

8.3. WEAK ECHO REGION (WER). While most weak storms consist of a single "pillar" of reflectivity, a weak echo region (WER) is an area of significant reflectivity suspended aloft in or near the updraft area. This suspended reflectivity is also referred to as *overhang*. The feature is typically found on the inflow flank of a severe thunderstorm. The weak echo region consists of precipitation falling out of the storm, but due to the wind profiles it encounters on its way down it is carried back into the storm area.

> **The vault**
>
> The vault has ambiguous meaning in severe weather forecasting and its use is avoided in this book. The echo free vault, or radar vault refers to the weak echo region. The visual vault refers to a vaulted area where cloud bases are much higher, and is often on the periphery of or within the downdraft region.

8.4. BOUNDED WEAK ECHO REGION (BWER). The bounded weak echo region (BWER) is a column of weak reflectivity which seems to punch upward into a WER or into the storm precipitation area itself. It may be identifiable using only one elevated reflectivity scan; it appears as an O- shaped weak echo area or a U-shaped weak echo area. Note that WER and BWER centers can contain light to moderate reflectivity. Often this is an unavoidable artifact due to "smearing" of the detected storm by beam spreading.

Finding the updraft is an important part of severe thunderstorm forecasting. If a BWER is present, the updraft can be presumed to be colocated with the BWER. If no BWER is present, the strongest reflectivities in the upper portions of the storm (at about 20,000 to 30,000 ft AGL) can be assumed to be immediately above the updraft, since such intense reflectivities can only be generated by the updraft (Lemon 1998).

When a WER or BWER exists, maximum echo tops are often over the top of the WER or BWER rather than over the main precipitation core. This can help confirm the BWER's existence and location.

The future of Doppler
One of the most critical problems that meteorologists have had to face has been to obtain reliable information concerning the actual existence of a tornado or funnel cloud in sufficient time to warn those in threatened areas. It is believed that Doppler radar would aid in easing this problem so that we can greatly improve our ability to prevent loss of life due to those storms.

R.L. SMITH & D.W. HOLMES
U.S. Weather Bureau, 1957

8.5. SUPERCELL COLLAPSE. As far back as 1965 it was noticed that some severe cells exhibited an apparent collapse phase where strong updraft signatures disappeared. Over the next few decades such collapses were linked to the onset of damage and tornadogenesis. Signals which suggest a collapse include the filling in of a BWER, loss of echo tops, decrease in VIL, descent of suspended high intensities, and increase in reflectivity in the lowest elevations. However, there has been considerable debate over the years about whether these signatures actually indicate an updraft collapse, and whether it may be ascribed to the collapse of a smaller constituent updraft or even to changes in pressure distribution or flow within the storm. In any case, collapse-like signatures are always an indicator of an important change in the storm and always warrant further interrogation of radar products.

8.6. STORM AHEAD OF A LINE. Since the 1960s it has been recognized that storms which form ahead of a squall line are candidates for severe weather or tornadic activity. This is not due to any special atmospheric processes but rather because any isolated, non-linear storm in an environment already favorable for organized convection is a prime candidate for severe weather. Storms ahead of a

Figure 8-9. Radar is not just for looking at storms. It can be used to refine the surface analysis, showing important boundaries as seen here. *(Top: Tim Vasquez, Digital Atmosphere; Bottom: Courtesy UCAR, weather. rap.ucar.edu)*

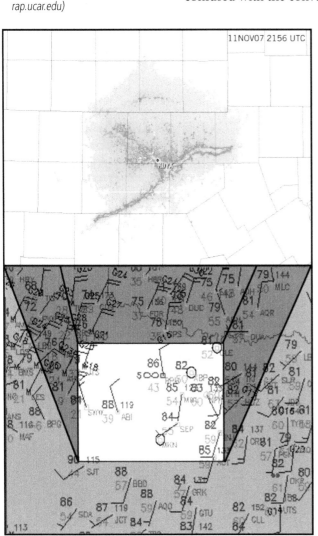

line can sometimes be anticipated by looking for boundaries in the warm sector on surface, satellite, and radar imagery.

8.7. INFLOW CONVERGENCE. While storm forecasters often talk about gate-to-gate shear and mesocyclone couplets, inflow convergence can sometimes be an important feature since it precedes tornado development. Inflow convergence is a convergence couplet signature at the lowest radar elevation and is in the inferred updraft area. Unfortunately it can only be resolved when it is relatively close to the radar. The velocity differential in the convergence couplet rarely exceeds 15 kt. Furthermore convergence can easily be confused with the convergence that occurs along outflow boundaries, though the latter tends to be linear in structure.

8.8. MESOCYCLONE ROTATION. Storm-scale rotation on the scale of several miles and persisting for about 10 minutes or more indicates a mesocyclone. This is indicated by a rotational couplet: a pair of maximum inbound/outbound velocities both at nearly the same range. The mesocyclone core diameter is the distance between the couplets, and the rotational velocity is the difference between the velocities observed in the couplets. The couplet may show both rotation and convergence signatures, especially when sampled at close range near the ground where convergence dominates rotation.

8.9. STORM TOP DIVERGENCE. This feature is often overlooked by forecasters but it provides a valuable indirect measure of updraft strength. The upper elevations corresponding to the anvil top are examined to find couplets of strong divergence, where the updraft has lost its momentum and is spreading horizontally. This indicates the summit of the updraft of the storm. The difference between the maximum velocities observed in the couplet yields the storm top divergence speed. Values of 50 to 75 kt are indicative of a severe thunderstorm.

8.10. BOW ECHO. The bow echo (Fujita 1978) is a radar reflectivity signature featuring a linear thunderstorm echo with a bow-shape. It usually points toward the axis of movement, and measures anywhere from 20 to 200 km in size. At small scales it is associated with downburst activity, and at large scales it is associated with derecho windstorms. Tornadoes are possible along the leading edge of the bow echo (mesovortex tornadoes) as well as at each end of the bow echo (bookend vortices). Details on the bow echo are found in Chapter 3, *Mesoscale Convective Systems*.

8.11. LINE ECHO WAVE PATTERN (LEWP). The line echo wave pattern was described by Nolen in 1959. The term is rarely used today since the term is descriptive in nature and bow echo storms and derechos are now recognized as distinct convective phenomena.

8.12. MID-ALTITUDE RADIAL CONVERGENCE (MARC). This is a radar-observable signature in an MCS that may precede the descent of a rear-inflow jet and the beginning of a bow echo. It is a spot of locally strong convergence about 3 to 5 km in diameter embedded within the larger-scale convergence on the storm's leading edge (i.e. ahead of the axis of highest reflectivity). Furthermore, it is found in the mid-tropospheric elevations, at about 3 to 9 km AGL. A velocity differential higher than 50 kt is a strong MARC indicator. Overall this feature identifies a potential location where the rear inflow jet impinges on the updraft. The greater the convergence, the higher the likelihood that basic conservation of mass will force the rear inflow jet to the surface.

8.13. REAR INFLOW NOTCH (RIN). The rear inflow notch is a bit of a misnomer as it is actually comprised of strong *outflow* from the storm. This is a region of unusually low reflectivity and enhanced velocity on the lowest horizontal radar elevation which appears to augment a mesocyclone embedded in the line or along its leading edge. It is comprised of descending (drying) air and appears to

---

Terminal Doppler Weather Radar (TDWR)

In the United States, the Federal Aviation Administration operates a network of about 45 special radars near major airports designed to allow controllers to be alerted to wind shear hazards, precipitation intensities, and wind shifts. The system was designed at MIT in the early 1990s and deployed starting in 1994. Data has not been available to the public because of difficulties bringing it onto weather datastreams, but by 1999 some TDWR reached the Internet and during 2009 all of the stations are expected to be placed online for general consumption.

The system differs slightly from the WSR-88D as it has a narrow (0.55 deg) beam width, strong ground clutter suppression, uses the C-band (5 cm) rather than S-band (10 cm), and completes a volume scan once per minute.

Access to TDWR has been limited to several sites due to delays in feeding the data to distribution lines. However during 2009 most of the sites became available for public access.

mark the arrival of downdraft air or a rear inflow jet segment at the surface. It may be preceded by a MARC signature (q.v.).

8.14. POLARIMETRIC HEAVY RAIN SIGNATURES. Polarimetric radar is excellent for differentiating situations where high reflectivity is due to hail (low rain rate) and due to water content (high rainfall rate). Cells showing high horizontal (regular) reflectivity and high levels of specific differential phase (K_{DP}) indicates very strong scattering from oblate particles, suggesting a cell with substantial rainfall production.

9. Hail signatures

9.1. INTENSE REFLECTIVITY. The traditional method for finding hail on radar is simply looking at the lowest elevation for reflectivities with very high intensities. This is predicated on the assumption that reflectivities above 60 dBZ are rarely associated with pure rain and involve some sort of ice, which is a much better reflector. However this correlation is marginal. For instance large, dry hailstones can be associated with moderate reflectivity values.

9.2. THREE BODY SCATTER SPIKE (TBSS). Few hailstorms exhibit a three body scatter spike (TBSS) signature, but when a TBSS is

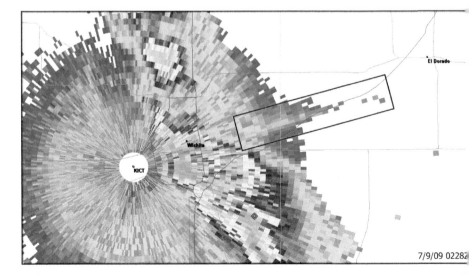

Figure 8-10. Three-body scatter spike extending from the storm away from the radar. When this occurs it indicates large hail is likely. This storm occurred 8 July 2009 at 9:28 pm CDT and produced large hail in Wichita, Kansas. *(Tim Vasquez / graphics GRLevel2 <grlevelx.com>)*

seen it is almost always a direct indicator of large hail. The TBSS is a false spike appearing on radar which seems to extend away from a hailstorm opposite the radar location. The TBSS is named after the three bodies which, together, scatter a single radar pulse: a group of hailstones, the ground, and the same group of hailstones again. Large, wet hailstones have extremely high radar reflectivity, enough to where energy can be reflected to the ground, back to the hailstones, back to the radar, and still be detectable. Because of the extra distance involved in this side trip, and because the radar by design assumes that all reflection is occurring along the beam axis, the radar display system paints a horizontal spike extending about 10 to 30 miles away from the storm. The TBSS usually has a weak reflectivity of less than 20 dBZ, and emanates from a highly reflective core of 60 dBZ or more. In some cases it may be more visible on spectrum width imagery than on reflectivity products. Solid hailstones without a coating of water will not produce a TBSS signature. The use of TBSS signatures can be especially helpful in high-shear supercell environments.

9.3. HAIL DETECTION ALGORITHM (HDA). The WSR-88D offers the HDA algorithm which attempts to identify hail locations and calculate three attributes of hail at each location. This WSR-88D

Figure 8-10. Radar beam attenuation behind a hail core about to drop giant hail on the Mayfest events in Fort Worth on 5 May 1995 2350 UTC. The core at this point showed a reflectivity of 73 dBZ. The attenuation sector (black lines) have been added manually here to emphasize the location. These attenuation sectors are especially evident on animation and may interfere with the interpretation of storms. This particular storm killed 13, injured 100, and produced $1 billion in damage, all from wind, hail, and lightning. *(Tim Vasquez / graphics GRLevel2 <grlevelx.com>)*

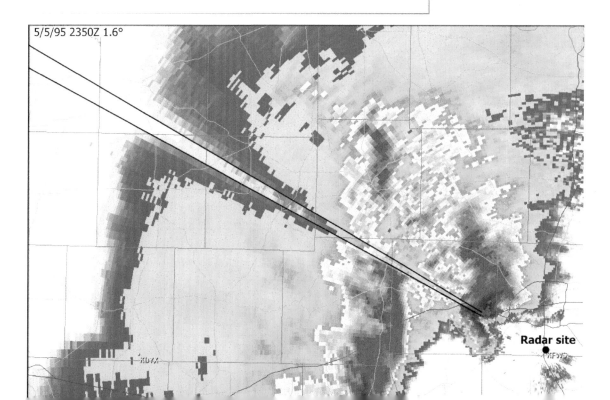

Figure 8-11. **The proximity of an outflow boundary** to the storm can be used to gauge storm character. The outflow boundary has moved quite far from the storms on the right, so they are considered outflow-dominant. The large storm on the left shows the outflow boundary very close to the cell, suggesting the storm is receiving organized inflow. In that respect it has the highest potential for supercell development or tornadogenesis. *(Tim Vasquez / graphics GRLevel2 <gr-levelx.com>)*

algorithm, its latest incarnation fielded in 1996, uses the height of the freezing level and the distribution of reflectivity near a storm center. It produces three quantities: probability of hail (POH), probability of severe hail (POSH), and maximum expected hail size (MEHS). Because the algorithms do not account for topography, they will underestimate the risk for elevations that are significantly higher than the radar and will overestimate it for points that are significantly lower than the radar.

9.4. HIGH SPECTRUM WIDTH. Bunkers and Lemon 2007 proposed that storms with the potential for very large hail might be identified by the presence of wide updrafts, as revealed by spectrum width products. These show as a spectrum width minima, indicating the updraft core. Since large hail favors broad updrafts, the width of this core is the key: an updraft width of 3-8 nm is considered to be ideal. However spectrum width products do tend to be noisy and difficult to interpret, and the updraft often does not show itself in the spectrum width data. The signature may also be masked by mesocyclone circulations and the effects of a TBSS.

9.5. STORM TOP DIVERGENCE. The estimated divergence of winds at the anvil level provides a measure of the updraft strength. The greater the divergence, the stronger the updraft. The divergence is the difference between the maximum inbound and outbound velocities in the couplet; values of over 75 knots are favorable for large hail. Unfortunately because it is a shallow pattern and may not be detectable due to beam width or excessive elevation, nothing should be inferred from a lack of storm top divergence.

9.6. WEAK ECHO REGION (WER). A very strong updraft will produce an elevated core that is largely devoid of precipitation particles compared to parts of the storm around it. As a result, it has weak reflectivity and is known as a weak echo region (WER) (Chisholm 1970). The weak echo region is often full of supercooled water droplets, and this provides a potent region for hailstone growth. Research has found that areas above a WER or BWER is where significant hailstone formation occurs. Giant hail is often associated with BWER signatures.

9.7. ELEVATED REFLECTIVITY CORE. A core of elevated precipitation that is higher than usual is an indicator of hail. In fact, reflectivities exceeding 55 dBZ at the -20°C level or higher are a very strong indicator of hail. Cross-sectional radar products are usually required to observe this.

9.8. VERTICALLY INTEGRATED LIQUID (VIL). During the early and mid 1990s, VIL, a summation of echo intensity through a vertical column, was widely used to identify cells with large hail. However it was proven to have no real correlation with hail events (Edwards and Thompson 1998). One valid reason for this is that VIL was designed to filter hail, and to accomplish this it truncates reflectivities that exceed 56 dBZ. Furthermore the VIL algorithm is range-sensitive: at close distances the storm falls within the cone of silence or at far distances, strong scatterers are not as well detected due to large beam width and this results in range biases.

9.9. VIL DENSITY. There has been some interest in using VIL density, which normalizes VIL for storm height by dividing VIL by the echo top in meters. This method is easy enough to calculate

Dual-pole quick reference

Differential reflectivity (Z_{DR})
<0   Conical ice
-1 to 0 Hail
0   Snow
2-4   Rain

Correlation coefficient (ρ_{hv}, CC)
0.96 to 1   Rain, dry snow
0.85 to 0.95 Mixed
<0.85   Lithometeors

Specific differential phase (K_{DP})
0   Hail
>0 Rain

Large hail and core height

One early technique (Lemon 1980) identified storms as producing large hail if the VIP 5 (50 dBZ) echo extends above 27,000 ft (Lemon 1980). This idea has been refined over the years to consider reflectivity height above the freezing level and is now a key principle in the hail detection algorithm used by the WSR-88D system.

The hook echo

The first hook echo was documented on 9 April 1953 on a storm near Champaign, Illinois (Stout and Huff 1953), and was inferred by Theodore Fujita to indicate storm rotation (Fujita 1958). A number of studies have been conducted which failed to find correlations between hooks and the presence of a tornado.

mentally while looking at radar products. VIL density gives greater weight to shallow storms with high VIL and weaker weight to tall storms with high VIL. A VIL density of 3.28 to 3.50 has been suggested as a threshold value for large hail. As with standard VIL, filtering of high reflectivities and range problems will affect the usefulness of VIL density.

9.10. POLARIMETRIC RADAR. Analysis of backscatter polarization shows tremendous promise for identifying the dominant type of precipitation particle in a specific range bin.

One polarimetric signature for hail is where Z_{DR} (differential reflectivity) values are close to zero. This indicates particles that are spherical and most likely to be hail, whereas raindrops have an oblate shape and show positive Z_{DR} values. This may appear as a "Z_{DR} hole" or "Z_{DR} core" of low Z_{DR} values embedded in a region of precipitation.

10. Tornado signatures

The greatest challenge to tornado detection is developing a four-dimensional mental understanding of the storm using the tools available, such as radar reflectivity and velocity, and making wise use of spotter and chaser reports from the field.

10.1. HOOK ECHO. The hook echo, or hook, is a radar signature which was first documented on a supercell near Champaign, Illinois in 1953. This fingerlike extension extends from the rear equatorward side of a storm and contains a cyclonic cusp. It has occasionally been referred to over the years as a "pendant echo".

A recent paper (Rasmussen et al 2006) proposed the term **supercell echo appendage**, owing to the fact that "hook echo" terminology fixates on the shape rather than the actual process that causes it, all of which may encompass a radar-observed life cycle of which a hook is only a small part. Hook echoes and supercell echo appendages all comprise mature stages of the descending reflectivity core (DRC), a three-dimensional signature described separately below.

The hook echo and its relation to tornadoes was documented as far back as 1953. It has also been known as a pendant echo. In dense precipitation areas, this hook echo may be so dense that only a notch of low reflectivities facing the inflow air is shown. The hook echo is considered to be a strong indicator of a mesocyclone. However the correlation with a tornado is weak.

For many decades, the hook echo was thought to originate from precipitation spiralling into the rotating updraft. However evidence also indicated that the hook echo developed vertically from descending precipitation in the rear-flank downdraft.

there is a growing body of evidence that at least in some storms precipitation descent within the rear-flank downdraft is a key cause.

10.2. TORNADIC VORTEX SIGNATURE (TVS). Tornadic rotation is primarily marked by extreme contrasts in speed between *two adjacent radials* within or near a suspected updraft area. This is sometimes called "gate-to-gate shear". Tornadic circulation should be suspected for any pixel-to-pixel velocity difference of at least 40 knots at the lowest elevation or 58 kt at any elevation. It is also important that this shear shows at least a rudimentary element of vertical consistency between other elevations.

This structure of intense gate-to-gate rotation with vertical consistency is called a tornadic vortex signature. The ability of the WSR-88D to detect a TVS exists only between about 20 and 70 nm. Beyond this range, the beam width masks all small-scale structure and is too high in the storm. Targets that are less than this range are masked by ground clutter and exhibit more of a rotational convergence signature than pure rotation.

It is important to remember that the presence of a mesocyclone greatly increases the probability that a TVS correlates to an actual tornado.

10.3. TORNADO DETECTION ALGORITHM (TDA). The WSR-88D tornado detection algorithm specifically looks for tornadic vortex signatures. It detects gate-to-gate shear of 49 kt or more in the lowest level, or 70 kt or more at any elevation.

Early Doppler radars

Though WSR-88D technology is popularly thought to date to around 1990, much of the groundwork had been laid decades earlier. The very first Doppler radar was a continuous-wave mobile unit deployed in Kansas in 1957 by the Weather Bureau. This yielded some of the first information about tornado wind speeds, suggesting winds of 100-200 mph but nowhere near the Mach 1 speeds that some theories suggested.

In order to plot meaningful information about the distribution of reflectivity and velocity along a beam, a Doppler radar needs to use pulses rather than a continuous wave. The vast amounts of signal processing needed to make this happen did not exist in the 1950s, but by 1964 such radars were developed at the National Severe Storms Laboratory in Norman OK and the Air Force Cambridge Research Laboratory in Massachusetts.

At NSSL in 1971 a full 10 cm Doppler surveillance radar was built, perhaps the first embryo of NEXRAD, but due to computer limitations, real-time processing of velocity was not available until 1977 with the development of the J-DOP system. By this time a number of universities had been experimenting with pulsed Doppler systems and reported great success. Finally in 1982 it was decided to create an operational network in the U.S. This became known as NEXRAD.

10.4. DESCENDING REFLECTIVITY CORE (DRC). An area of high-reflectivity underneath the rear overhang that descends to the surface near the rear-flank downdraft is known as a descending reflectivity core (Rasmussen et al 2006). In some cases it has been found that this DRC is the formative stage of the hook echo. The DRC takes about 5 to 15 minutes to descend to near the surface, where its arrival is speculated to be associated with tornadogenesis. In the 2006 paper all of Rasmussen's tornado samples were associated with a DRC, but many nontornadic supercells also had DRC signatures.

Research from the 1960s led to a popular notion that tornadogenesis was associated with an apparent collapse of the updraft, manifested by radar-observable signatures such as the filling-in of a bounded weak echo region, the descent of high reflectivities toward the surface, and a decrease in echo top height. It is likely that this "collapse" may be related to the concept of the descending reflectivity core.

10.5. POLARIMETRIC TORNADO SIGNATURES. Near the vortex, the radar may show dust and debris in the air. Since these have very different polarimetric characteristics, this can be useful for locating and confirming a tornado. Where moderate to high reflectivity is

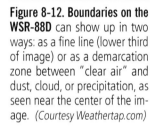

Figure 8-12. Boundaries on the WSR-88D can show up in two ways: as a fine line (lower third of image) or as a demarcation zone between "clear air" and dust, cloud, or precipitation, as seen near the center of the image. *(Courtesy Weathertap.com)*

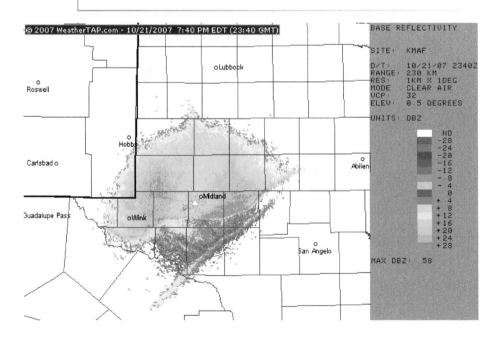

detected at a TVS location, check the characteristics of the polarimetric products. Differential reflectivity values will show values near zero, and correlation coefficients will be very low.

11. Other valuable signatures

11.1. FINE LINES. Fine lines, which are filament-like lines of low reflectivity, are frequently seen on radar imagery. They are caused by refraction along the boundary or the formation of cloud droplets from lift. Fine lines are extremely valuable for finding wind shift lines and fronts. In any situation where convection is expected to be surface-based, forecast target areas should be thoroughly evaluated for fine lines to try to locate all boundaries and resolve them with surface and satellite data. ¤

9
SATELLITE

Much of the information gathered [from a weather satellite] would be fairly meaningless to the average citizen. It would, however, tell scientists a great deal. From it, for example, they would acquire new facts about the conditions that influence the weather. The important thing to remember is that although the announcements for the plans for the development of an earth satellite is a somewhat startling one to the average person, to the scientists it is simply a logical next step in their researches.

— Dr. Alan T. Waterman, 1955
chief scientific advisor to satellite program

Satellite data is arguably one of the most important tools in pre-convective storm forecasting. Surface observation networks may detect the height and coverage of cloud forms and radar may hint at their presence, but only satellite imagery reveals its character. Quite often, just by inspecting visible satellite imagery a skilled forecaster can accurately isolate the location of thunderstorm development as much as an hour before the first radar echoes appear.

1. Image types

Most satellite data used in severe storm forecasting originates from geostationary satellites, which are semipermanently fixed 22,240 miles (35,790 km) above a given location on the equator. The United States constellation of satellites is known as the GOES (Geostationary Operational Environmental Satellite) network and covers the western hemisphere and adjoining ocean waters. GOES has been in operation since 1974. Many weather services also make use of polar orbiting or sun-synchronous satellites, but these are not often used in storm forecasting since images only arrive twice a day and must be georeferenced. Their main value to forecasting is in polar regions where the earth's surface is oblique to the satellite.

Most weather satellites produce three key types of imagery: *visible*, *infrared*, and *water vapor*. By combining various spectral channels we can also obtain *multispectral* imagery, which is useful for certain types of work such as fog detection and icing risk.

1.1. VISIBLE. Visible imagery detects clouds in the same color bands that the naked eye perceives, so interpretation is straightforward. Brightness corresponds to albedo, the reflectivity of light from the sun. The highest albedo values come from fresh snow and cumuliform clouds and appear white, while lower values are produced by cirriform clouds, fog, stratus, and ground and appear more grayish. Visible imagery is never available at night, though for decades the military DMSP weather satellites have allowed forecasters to view coarse nighttime visible imagery by moonlight, providing a superior way to detect fog and stratus.

In North America, severe storm forecasting flourishes due to the widespread availability of near-realtime high-quality imagery. Elsewhere in the world, most publically-released geostationary satellite data is often poor due to withholding of finescale imagery for commercial or bandwidth reasons, but this is changing. In Europe, for example, EUMETSAT recently began publicly distributing high-resolution visible images known as HRV (high-resolution visible) imagery.

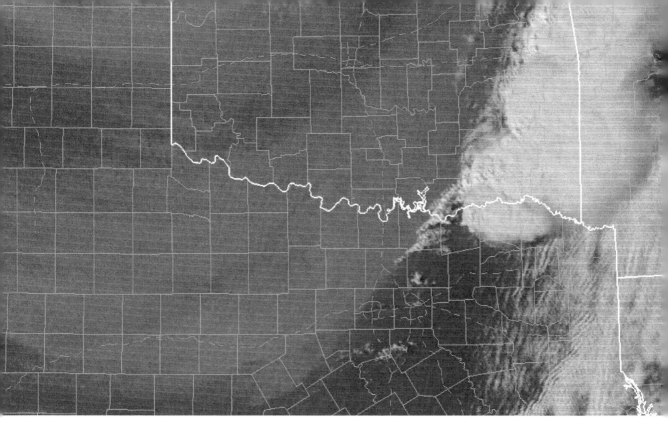

Figure 9-1 Supercells near Paris TX and Broken Bow OK on 2 April 1982 at 2130Z. This tail end storm along the Red River produced Oklahoma's only F5 tornado after 1976 until the May 1999 outbreak occurred. The synoptic-scale environment was dominated by very strong forcing and shear from a powerful storm crossing the Great Plains. It lofted large amounts of dust from west Texas eastward, visible as cloudy streaks in the left half of the image. Wichita Falls reported wind gusts to 51 kt and Lubbock to 54 kt. The banding on this image is an original artifact of the primitive sensors used 30 years ago, and was common on many images in that era. *(Tim Vasquez / graphics McIDAS)*

1.2. INFRARED. By sensing imagery in the infrared spectrum it is possible to receive detailed cloud imagery 24 hours a day. The key to infrared imagery is areas with low radiance are considered to be cold and are colored bright, while areas with high radiance indicate warm temperatures and are colored dark. This gives a fairly remarkable facsimile of visible imagery and at times it can indeed be difficult to tell the two apart. There are some exceptions, though; particularly when ground temperature and cloud temperature are equal and cannot be differentiated. This often happens in Canada in the winter where cold ground and cirrus look alike, and in the southern U.S. in the warm season where warm ground and stratus and fog appear the same.

1.3. WATER VAPOR. Water vapor imagery is a specialized product that uses bright coloring to identify areas of high specific humidity, though effectively it is only sensitive to upper-level moisture. The water vapor product measures the upwelling of radiation from the ground and atmosphere into space. Cold (bright, moist) areas imply either that upwelled radiation in a warm area was absorbed by water vapor, leaving little of it to escape into space, or that the location is in fact a cold area and no radiation existed in the first place.

Warm (dark, dry) areas are more straightforward, implying radiation in a warm area existed and escaped into space without being absorbed by water vapor.

Water vapor imagery is effective at detecting areas of potential instability. A potentially unstable atmosphere is dry aloft, and very dark areas are highly suggestive of dry upper-level conditions. Sustained vertical motion in these areas has the potential to destabilize the atmosphere and prepare the region for convective development.

Localized spots of darkening may indicate intense, mesoscale quasi-geostrophic disturbances that are moving through the flow and deserve closer scrutiny. The dark areas are associated with subsidence and are often paired with a corresponding area of strong upward motion. While this will not cause thunderstorms by itself, it can be enough to destabilize the atmosphere and perhaps remove enough of the cap to allow initiation.

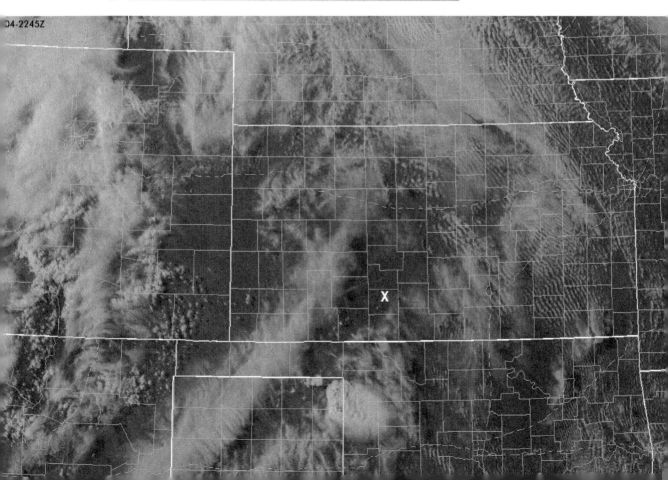

Figure 9-2 Cirrus over a target area can significantly degrade the use of visible imagery. In this example there are fortunately some gaps and it is probably possible to find some boundaries with the judicious use of surface and radar data. This example is for 2245Z on 4 May 2007, the date of the Greensburg Kansas F5 tornado (small "x") which hit a few hours later. *(Tim Vasquez / graphics McIDAS)*

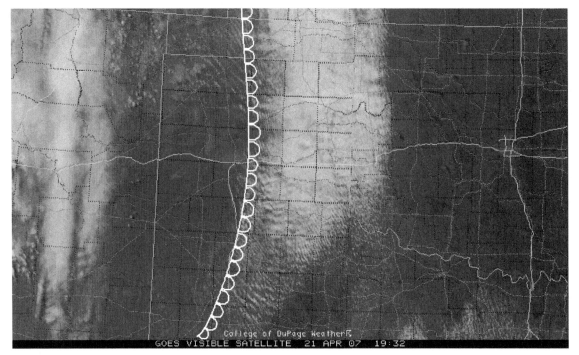

Figure 9-3. Cumulus and stratocumulus ahead of a dryline (scalloped white line). The broken to overcast cloud layer consists largely of closed-cell stratocumulus associated with strong capping or deep moisture. The open-cell stratocumulus and cumulus along the western and especially the southwest periphery is more indicative of stronger heating and mixing. Also note the presence of convective elements west of the dryline: stable rolls just behind it giving way to more vertical, "speckled" towers further west in closer proximity to upper level lift (suggested by the dense cirrus). Cumulus fields behind the dryline can become surprise initiation areas. *(Courtesy College of DuPage)*

2. Cloud forms

In visible imagery, cloud forms are distinguished primarily by texture and brightness. Before storms develop, severe weather forecasters primarily concern theirselves with low-level cloud forms since this yields information on low-level moisture and boundaries. Middle and upper level clouds are considered, too, but more in terms of finding upper-level disturbances as well as effects on surface heating.

2.1. STRATOCUMULUS. Stratocumulus is a type of cumuliform cloud with primarily horizontal development. Because of this, it tends to congeal into broken or overcast masses. They may be formed by stratification of moisture in the presence of lift, by the spreading of convective elements beneath an inversion, or both.

In some cases, stratocumulus may form into cells measuring 20 to 50 km in diameter. The closed cell pattern contains cloudy air at the center and clear air at the edges, and is most common near deep extratropical cyclones in regions of intense cold air advection. The

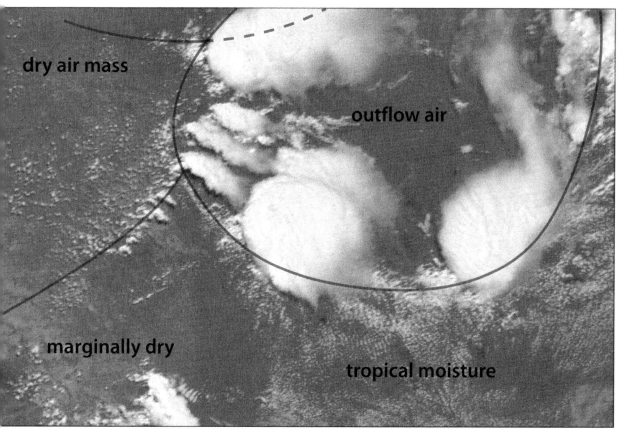

Figure 9-4. Outflow boundaries as seen on visible satellite imagery. Note that the primary indicator here is not so much a linear feature but rather a difference in low cloud type. The air mass with tropical moisture is composed mainly of cumulus streets, but the air mass with outflow air is clear except for cirrus debris. This example is from 31 May 1995; the storm in the center produced a destructive hailstorm in San Angelo, Texas about an hour later this image was taken. *(Tim Vasquez / graphics McIDAS)*

open cell pattern contains clear air at the center and clouds at the edges, and is associated with unstable patterns and may consist of cumulus and towering cumulus clouds.

If enough shear is present, a closed-cell stratocumulus field may organize into *transverse rolls*, with axes *perpendicular* to the cloud layer shear vector. For example if cloud base winds are calm and cloud top winds are southerly, these bands will be aligned west to east. The wavelength is about 5 to 10 km, large enough to be seen on satellite imagery. It is the author's experience that extensive transverse rolls late in the afternoon are associated with a cap strong enough to suppress all further development.

2.2. CUMULUS. In a relatively clear atmosphere just before thunderstorms develop, cumulus in the target area often

Figure 9-5. A warm front (dashed line) separates cumulus cloud streets (lower right) from flat stratocumulus with transverse rolls (upper right) in cool, stable air. This example was captured in northwest Kansas on 22 May 2008. A tornado hit the area near the circle. Even with the front well to the north, the cloud lines south may have had sufficient residual baroclinicity along their lengths from the departing front to augment tornado production. *(NASA/GSFC)*

Very often in warm, sunny conditions, cumulus clouds may form into longitudinal rolls known as *cloud streets*. The bands have an axis of orientation *parallel* to the cloud layer shear vector. For example if cloud base winds are calm and cloud top winds are southerly, cloud streets will exhibit a north-south alignment. The bands have a wavelength between streets of 2 to 10 km. This is extremely common in a benign summertime air mass.

The streets are created by the development of convection into counterrotating horizontal vortices, each one rotating in the opposite direction of its neighbors. The vortices either combine to produce upward motion, creating the visible cloud streets, or combine to produce downward motion, producing clear air. Streets are favored in unstable layers when speed shear without directional shear exists in the cloud layer, or occasionally when directional shear oc-

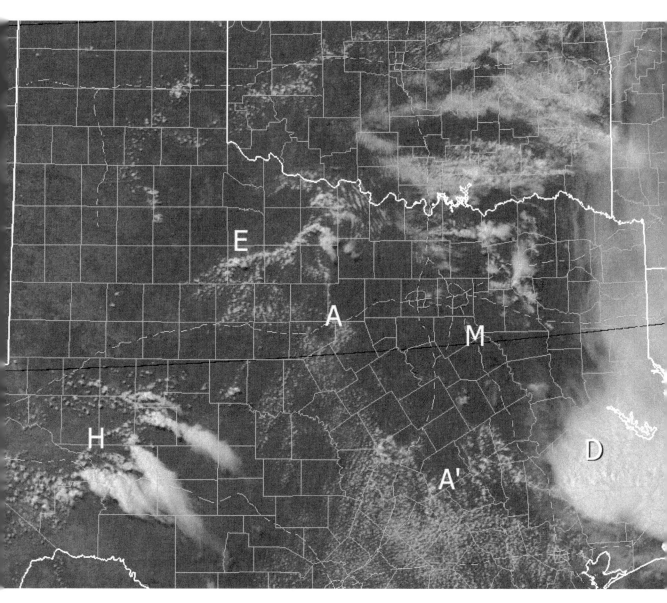

Figure 9-6. Pre-initiation cloud features as seen midafternoon at 2000 UTC on 29 May 1994. Note the decaying thunderstorm area (D) which had been moving southeastward during the day, leaving behind a clear outflow air mass (M). The southwest edge of this forms an outflow boundary which can be seen roughly from A to A', separating the outflow air from convectively unstable air to the southwest. There is an area of enhanced cumulus (E) which rapidly developed over the next hour into large thunderstorms which produced an F3 tornado at Newcastle, TX and another that nearly destroyed Cross Plains, TX. The outflow boundary A-A' was a significant contributor to the rotation in these storms. Higher based storms (H) are in the Pecos Valley region but were in an area of weaker low-level moisture, showing larger elements and marginal anvils. *(Tim Vasquez / graphics McIDAS)*

> **NSSFC and the MCV**
>
> At the NSSFC in Kansas City, a group of satellite weather meteorologists began working with us in the mid or late 1970s. They were showing us things from the satellite pictures that could be helpful for our severe weather forecasting. They began to notice the development of small scale low pressure systems (MCVs) on the back side of MCS systems and showed us where these were located. However, at that time we did not think that convective storms could produce a low pressure system. I remember one of our forecasters laughing about these satellite meteorologists telling us that there were MCVs next to an MCS.
>
> He thought that anytime there was an MCS moving around, a satellite meteorologist would always insist a low pressure system (MCV) was there when really it was just a convective storm system (MCS). Of course, sometime after the 1970s severe weather forecasters began to learn that in addition to synoptic scale upper troughs and low pressure systems, storm systems (MCSs) could also affect severe convective storm development and evolution. And within the last decade or so, severe weather forecasters have been able to make some accurate severe weather forecasts based on the MCS and MVC effects on severe weather development and evolution.
>
> BOB JOHNS, 2009
> former NSSFC (SELS) forecaster
> personal communication

curs without speed shear. Cloud models have been quite good at reproducing these patterns.

Some research suggests that these cloud bands produce small-scale axes of enhanced moisture. Gravity waves associated with anvils from MCS activity upstream may couple with these moisture axes to produce discrete initiation ahead of the MCS. This is still an ongoing area of study and the forecast implications are not clear.

2.3. OUTFLOW BOUNDARIES. It is extremely important for forecasters to scrutinize the finest-scale visible and infrared imagery for low-cloud lines and correlate them with surface and radar data. These may highlight outflow boundaries, fronts, and wind shift lines that are associated with convergence and enhanced shear. Even if an obvious boundary is not found, it may be demarcated by an abrupt change in the *character* of clouds across the zone. An outflow or postfrontal air mass will tend to be relatively devoid of low clouds, or may contain stratified forms such as stratus and stratocumulus.

2.4. ENHANCED CUMULUS. When deep convection is imminent, cumulus clouds tend to aggregate into clusters of vertically-developed clouds. Satellite imagery shows localized, persistent areas of distinct, bright tops. This pattern is often referred to by forecasters as *enhanced cumulus*. At this time observers in the field will see moderate cumulus, towering cumulus, or young unglaciated cumulonimbus clouds.

3. Storm signatures

3.1. OVERSHOOTING TOP. The overshooting top of a cumulonimbus cloud often appears on both visible and infrared imagery. On visible imagery, the albedo of the cumulonimbus tower is much brighter than that of anvil material. It may also exhibit a great deal of diffuse reflection on the sunward side and shadowing on the other. On infrared imagery, the temperature is much colder (brighter) than that of the anvil, producing a very localized cold spot. In general, the more prominent the over-

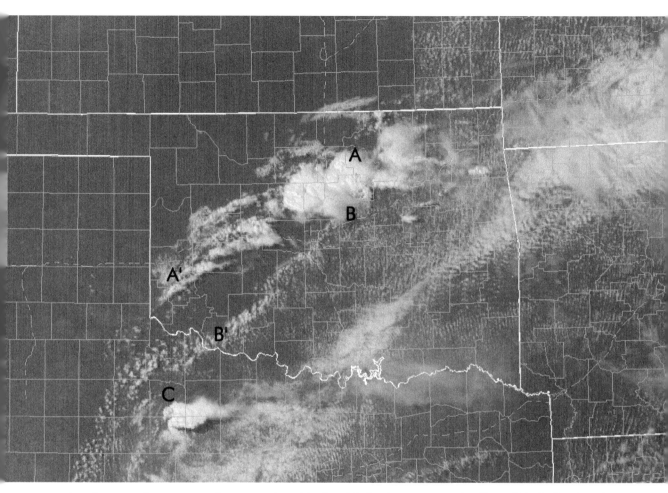

shooting top, the stronger the storm, but this should also be assessed against sounding data since a weak tropopause inversion may allow weak updrafts to penetrate to unusually large heights.

3.2. ANVIL SHADOW BAROCLINICITY. As described in the chapter on tornadoes, anvil shadow baroclinicity is suspected to a factor that augments the production of horizontal streamwise vorticity. If a distinct anvil shadow is noted, and there is fairly intense solar heating, and also if the storm is tracking roughly in the direction of the shadow edge, then this storm has an enhanced potential for rotating updrafts and tornadoes.

Figure 9-7. Outflow boundaries. On this 2045Z satellite image, many forecasters would show concern about the line of storms at A-A', however these are high based. Of more concern is the outflow boundary B-B', which produced two large supercells in central and southwest Oklahoma. Oddly enough the storm at C died a couple of hours later, but was not on the boundary. This image is for 9 May 2003. *(Tim Vasquez / graphics McIDAS)*

3.3. ENHANCED-V. The enhanced-V, formally described by McCann in 1983, is an *infrared* satellite signature on the top of an anvil cloud. It consists of a V-shaped area of cold tops seen on enhanced infrared imagery around an updraft or overshoot. The pointed end faces upwind. It is a localized feature with a dimension of only about 10 or 20 km, so fine-scale infrared imagery is required.

The enhanced-V is caused by barrier flow around an updraft, which spreads cold cloud material in a V- or U-shape around the overshoot and its downstream area of warmer wake subsidence (referred to as a CWA, a close-in warm area). During the era when this signature was studied, operational Doppler interrogation was nonexistent in the United States and satellite imagery played a pivotal role in daily forecasting. The feature is still of great value in assessing storm character in locations where Doppler radar data does not exist. A recent verification study of enhanced-V occurrences (Brunner et al 2007) found that about 90% of the time they were associated with severe weather occurrences, with hail being the most likely result. ¤

10

DIAGNOSIS

The plan which I would suggest is that each Signal Service observer be furnished with a tabulated statement of the weather, as made up at Washington from the telegraphic reports of the 7 a.m., 3 p.m., and 10 p.m. observations. From this he can quickly make his own synoptic chart showing the exact state of the weather over a large extent of country. The observer having carefully studied this chart, and having learned the local causes which may influence the weather at his own station, is far better able to predict the weather for his own locality than the Signal Service is or ever can be ... Combining this [idea] with local observations I soon found that the accuracy of my predictions rapidly increased, and if I was unable to obtain this daily table of observations I would give up all attempt at weather prediction.

— J. Bradford Sargent, 1887
American Meteorological Journal

Prognosis synthesizes all the available data: surface, radar, satellite, upper air, and even model forecasts, and also makes use of conceptual models. The forecaster reviews all of these elements and considers ingredient-based and pattern-based forecast methods, essentially identifying whether storms are possible and where they will occur.

The large scale environment shapes the day-to-day nature of severe weather patterns. While its mechanisms on the whole do not actually trigger thunderstorms, it plays a major part in preparing and shaping the pre-storm environment. Large-scale weather patterns advect substantial amounts of sensible and latent heat from one place to another and are powerful influences on the stability or instability of the atmosphere. All of this greatly influences the character of storms that do develop.

1. Upper level patterns

As was recommended in Snellman's 1982 forecast funnel, the analysis and diagnosis should progress from large-scale to small-scale, considering first the context for severe weather and then focusing on the severe weather mechanisms theirselves. Upper-level conditions are analyzed by the forecaster in terms of temperature, wind, pressure, and humidity, the latter typically expressed as relative humidity, mixing ratio, or dewpoint.

For novices, the system of pressure is somewhat unusual: instead of measuring the pressure at a specific height, we measure the height at a specific pressure level. This height is technically not geometric height above mean sea level but is actually *geopotential height*. The two are very similar but geopotential height accounts for the slight change in gravity with height and latitude, which used in equations of motion simplifies calculations, and at the surface is equivalent to geometric height, but in the upper troposphere is about 0.2% lower than geometric height in temperate latitudes. High height is equivalent to high pressure and low height is equivalent to low pressure.

1.1. PLANETARY BOUNDARY LAYER (PBL). The planetary boundary layer is that layer in contact with the earth's surface which contains convective cells and turbulent mixing, including turbulence

Analysis and diagnosis

If, as we have indicated, the careful analysis of surface observations is important to the task of forecasting, then doing a detailed surface analysis and diagnosis is an indispensable part of forecasting. Time simply must be found to accomplish this end. This task is particularly pertinent to the problem of forecasting deep convection. We believe it is not feasible to attempt forecasting convection without a commitment to do regular, detailed surface analysis designed to depict the pertinent structures and their evolution. The evolution is especially important, so unquestioning adherence to "continuity" may blind the analyst to important changes in structure occurring during short intervals.

FRED SANDERS &
CHARLES DOSWELL III
*A Case for Detailed
Surface Analysis*, 1995

> **The Guadalupe Pass rule**
>
> One pattern recognition rule noted by forecasters in the 1980s uses the post-dryline winds at Guadalupe Pass (KGDP), Texas, to "point" to where development might be expected along the Texas-Oklahoma dryline. This is loosely related to where the strongest synoptic-scale convergence is coming together.

due to friction with the ground. In an ideal atmosphere, horizontal winds are driven only by the balance of forces between pressure gradient, Coriolis force, and thus they are in geostrophic balance. Convection and turbulent mixing, when introduced, adds frictional forces to the motion of air parcels within that layer. This disrupts wind fields, slowing them and turning them more directly toward low pressure due to the decrease of the velocity term in the Coriolis force equation. Consequently, winds within the PBL are largely out of geostrophic balance while those above are in balance — and where they aren't it is likely significant processes are occurring.

The PBL is confined to the surface underneath inversions, including caps and radiational inversions at night. During the day with solar heating, the PBL grows upward many kilometers in height. In desert regions and behind drylines, it may reach through most of the troposphere. The forecaster must be aware of the vertical extent of the PBL on upper air charts: for example, 700 mb charts in the Rockies will often intersect the PBL during the summer, complicating comparisons with stations in Kansas and Nebraska. A check of soundings and familiarity with the site location can help identify the vertical extent of the PBL.

1.2. LONG WAVES. The upper-level patterns are dominated by low heights at the poles and high heights in tropical regions. The polar lows extend equatorward at various locations in the form of troughs. In between these troughs are ridges. Troughs and ridges with dimensions of synoptic or global size are called long waves. The root cause of low heights aloft are cold air in the layers below. This cold air is dense and tends to accumulate close to the ground, concentrating mass in the lower levels and leaving a void aloft.

1.3. SHORT WAVES. Waves in the atmosphere can occur on many different scales. At a smaller scale than the long wave is the short wave, which is synoptic in size. It represents a baroclinic instability in the atmosphere known as a quasi-geostrophic disturbance.

The presence of a short wave has an effect on the growth of upper-level troughs. When low-level cyclones grow and deepen, their circulation causes cold air to move equatorward on the west side. Consequently, upper-level heights fall west of the surface low and this reinforces the short wave trough. There is also a corresponding

DIAGNOSIS   205

area of height rises ahead of the surface low. Because of this relationship of the surface low with the short wave pattern, we can assume that areas on the upper-level charts downstream from a short wave trough but upstream from a short wave ridge are associated with cyclone development.

Interestingly, the amplification of the upper-level patterns in this manner helps develop and deepen the surface low. This is the process of baroclinic self-development. For this reason, baroclinicity, the existence of a temperature gradient, is considered an important precursor for system development.

As the low-level cyclone matures, it eventually occludes, since the cold air flows completely around the low and physically separates the surface low from the warm air. Since the low is no longer associated with a temperature contrast, it is no longer associated with an upper-level trough and ridge. Rather, the cold air filling the low causes the upper-level trough to close off into an upper-level low directly above the surface low. This is called a "cold core low" or a "decaying low". It may remain intense for a day or two but will gradually migrate poleward and dissipate. In North America, this graveyard is usually found in the Hudson Bay and around Baffin Island.

1.4. MESOSCALE DISTURBANCES. Severe thunderstorm forecasters frequently encounter small-scale waves in the upper-level flow which are barely resolvable with the given observational data density. They may be spotted on satellite imagery and profiler data. They may bring intense, locally-concentrated lift to a target area and be poorly resolved by numerical models. These disturbances are transitory and tend to move with the prevailing winds aloft. Since they are poorly detected by the upper-air network, they are

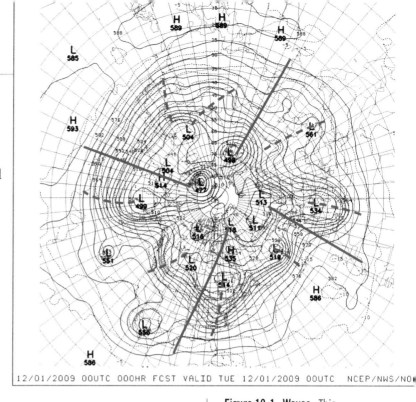

Figure 10-1. Waves. This chart shows an example of numerous mid-tropospheric waves circling the northern hemisphere on 1 December 2009. This chart uses the 500 mb level, which is high enough to vaguely make out the long wave pattern (the four thick lines) but low enough to where they are largely masked by short waves (dashed thick lines). And embedded in this flow are smaller-scale waves that are barely resolvable even with closely-spaced radiosonde stations, profiler data, and satellite imagery. *(Tim Vasquez / graphics NCEP)*

Figure 10-2. On the Great Plains, 700 mb temperature has traditionally been used as a dead-reckoning method for assessing cap strength, primarily when no other data is easily available. The threshold figure is dependent on the season. In the graph below, the shaded area indicates that storms are possible, with the white area suggesting the atmosphere is too capped. It should be regarded as a climatological technique and it makes assumptions about upper air conditions which might be regarded as common but at the same time not existent on a given day. *(Tim Vasquez)*

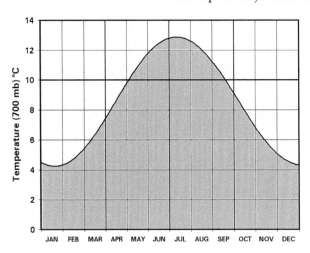

best tracked with mesoscale surface isallobaric analysis and satellite imagery.

1.5. JETS. A jet, which is any longitudinal core of strong winds, is caused by a strong height (pressure) gradient at that level. In turn, a height gradient aloft is caused by air masses below which have markedly different densities. Therefore it follows that jets are closely linked to where low-level and mid-level temperature contrasts, i.e. fronts, are found. Jets which are specifically associated with the polar front, the division between tropical and polar air, are known as polar front jets (PFJ). Another type of jet called the subtropical jet (STJ) is sometimes seen in the southern U.S. and other subtropical locations; it is not associated with low-level fronts and tends to be seen only at the highest levels of the troposphere.

As early as the 1950s it was recognized that storms tend to form where the upper-level jet crosses the moist sector. Furthermore, baroclinic lows, troughs, and other disturbances are found underneath the jet and embedded in this flow, providing numerous locations for producing sustained surface convergence for initiation and backing surface winds to augment rotating storms.

1.6. FOUR-QUADRANT CONCEPT. The conceptual model of a jet streak ("jet max") in the atmosphere emphasizes two quadrants of rising motion and two quadrants of sinking motion. Standing in a standard frame of reference with one's back to the upper-level wind and upon the jet streak itself, upward motion is favored in the left front quadrant and right rear quadrant, and downward motion in the right front quadrant and left rear quadrant. All four of these quadrants are also referred to as left exit, right entrance, right exit, and left entrance to illustrate the entrance and exit of air into these quadrants.

The mechanics of how these quadrants work is beyond the scope of this book, but in short the quadrants are caused by convergence and divergence of air in the upper troposphere which enters various portions of the jet streak pattern and become imbalanced. Their existence has been underscored in numerical models (Moore et al 1990). The convergence and divergence occurs as the air tries to restore the balance of forces. Convergence aloft produces downward motion and divergence aloft produces upward motion.

Severe thunderstorms have long been theorized to favor the left front and right rear quadrant of a jet (Beebe and Bates 1955). Recent studies indeed have found a positive correlation (Rose et al 2004). The divergent quadrants probably act to enhance severe weather activity by augmenting pressure falls, which can locally increase shear, and produce large-scale erosion of the capping inversion.

1.7. LOW-LEVEL JET (LLJ). The low-level jet is a mesoscale or small-scale synoptic jet in the lower troposphere, with a strength of about 30 to 50 kt and a core height of about 0.5 to 1 km. It is important in advecting richer moisture or warmer temperatures into a threat area, thereby destabilizing it. In the U.S., the LLJ is generally superimposed on the southerly flow of tropical moisture from the Gulf, and is strengthened in advance of an approaching upper-level jet, the so-called *exit region*, and effectively links the low-level subsidence in the right front quadrant with the rising motion in the left front quadrant of this upper level jet. The jet also has an effect on the hodograph, stretching the lower portions and intensifying speed shear and helicity.

At night, the LLJ strengthens as radiational cooling takes place at the ground, producing a thermal inversion that "uncouples" the free atmosphere from friction and turbulent processes near the ground. On the 1200 UTC soundings in the U.S. it is observed at the levels around 925 to 850 mb, and higher in the high Plains.

2. Moisture

In previous chapters we have established the principles by which moisture generates thunderstorms. But where does this moisture come from? The atmosphere has no ability to materialize moisture out of thin air. Moisture has to actually be transported from its source and placed within the pre-storm environment. Sources of evaporated moisture include evaporation from water

Divergence aloft

During the 1960s and 1970s, upper level divergence was considered to be an important predictor of severe weather. While this can be evaluated by machine analysis, diffluence (the spreading of contours) has often been depended upon to infer divergence. In nearly all instances diffluence (lateral divergence) is cancelled out by speed convergence (longitudinal convergence).

Also it must be remembered that upper-tropospheric divergence is simply a means of inferring synoptic-scale ascent, which does not trigger storms but destabilizes the tropospheric column of air over time, increasing the lapse rate and weakening any cap that might be in place.

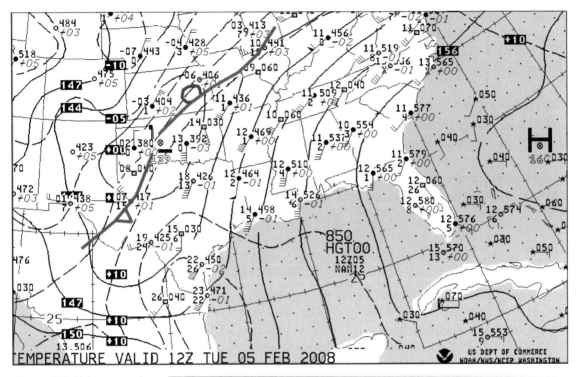

Figure 10-3. Moisture can be assessed on the 850 mb chart, not just the surface chart and sounding. Here we see conditions on the morning of the 2008 Super Tuesday tornado outbreak in Tennessee, Kentucky, and adjoining states. It reveals dewpoints of 46 to 52 deg F at 850 mb throughout the threat area. Note that moisture values on the station plots are dewpoint depression, which is inversely proportional to relative humidity and is more tied to the presence of cloud material (5 deg C or less indicates a high likelihood of cloud). To obtain dewpoint, the dewpoint depression must be subtracted from the temperature. *(Tim Vasquez / graphics NCEP)*

bodies like the Gulf of Mexico, evaporation from precipitation on the ground, and evapotranspiration from irrigated farming belts and forested areas.

2.1. GULF MOISTURE. During the winter and spring, the key source of rich moisture on the Great Plains is tropical air from the Caribbean and Gulf of Mexico region. However, during the summer and fall a significant source of rich moisture is stagnant air masses over the southeast U.S. that have long dwell times over forested and irrigated regions and continuously gain moisture through evapotranspiration.

When synoptic-scale onshore flow occurs on the Gulf Coast, accompanied by moisture advection, this is said to be a *return flow episode*. In weak return flow episodes, the maximum depth of the tropical moisture is about 0.5 to 1 km, while in strong return flow episodes it can deepen to 2 to 3 km. During the warm season, even greater depths are common, sometimes encompassing the entire troposphere. The tropical air mass is surface based, though if cold air has been in place across the region, a warm front will be in place with isentropic ascent of the tropical air mass, that is until the cold air has been mixed out or recedes poleward.

One important indicator of moisture richness (quantity and depth) is the character of cloud cover. Shallow moisture is represented by clear skies throughout the day or a layer of stratus that rapidly dissipates or turns into fair weather cumulus. Deep moisture is indicated by overcast stratus and stratocumulus clouds in the morning that give way to broken or overcast stratocumulus throughout the afternoon. A common rule of thumb in significant tornado outbreaks is that rich low-level moisture should exist over the forecast area for at least 24 hours, partly to demonstrate that sufficient moisture richness and depth has been achieved for supercell development.

2.2. EVAPOTRANSPIRATION. Evapotranspiration, the addition of water vapor to the air by plant life, is a significant process in forested and agricultural regions and is especially significant in corn production areas during the warm months. At the synoptic scale, it can add appreciable moisture to an air mass. In the northern Great Plains, evapotranspiration is frequently more significant in sustaining or raising dewpoints than advection from the Gulf of Mexico.

Dewpoints near an irrigated field can locally be increased by 10°F or more if there is little mixing of the boundary layer, and can result in dewpoints exceeding 80°F. This effect is quite common in corn production regions during the summer. But it is important to remember that a surface dewpoint reading is a surface measurement and this can lead to false assumptions about its depth. A very strong dewpoint anomaly is often correlated with shallow and poorly-mixed moisture confined within a stable layer near the ground. Using a sounding parcel which has been affected by evapotranspiration, without any common-sense corrections, can produce unrepresentative results.

Historically, forecast models have had considerable difficulty with evapotranspiration. In agricultural regions model bias may become apparent, especially during the growing season, in areas where moisture sources are not well parameterized.

Evapotranspiration

Large areas of crop fields, most importantly the corn fields in the Midwest, are an important source of atmospheric moisture during the warm season. They can cause unusually high dewpoint readings which are indeed accurate, although this enhanced moisture exists only in a shallow layer and may not have much impact on a mixed boundary layer.

2.3. MOISTURE CONVERGENCE. The quantity called moisture convergence has long been associated with thunderstorm initiation areas. Known properly as *moisture flux convergence* (*MFC*), it is defined as the pure convergence of the wind field modulated by moisture, q, and considering changes due to advection:

$$\text{MFC} = -u\frac{\partial q}{\partial x} - v\frac{\partial q}{\partial y} - q\left(\frac{\partial u}{\partial x} + \frac{\partial v}{\partial y}\right)$$

Moisture convergence essentially attempts to predicts the change in moisture over time at a given location. It is based on the shaky assumption that air movement at this level is purely horizontal and there are no diabatic processes taking place like precipitation or evaporation. The assumption is that high values of MFC predict where moisture will increase and parcels will become more buoyant.

In practice, MFC is a simplification of what really takes place in the atmosphere. Like any diagnostic parameter that examines one level only, it fails to consider influences throughout the atmosphere. It does not take into account the effects of stabilization or destabilization aloft, entrainment of dry air aloft, mixing, surface evaporation, convergence aloft, and other factors. Furthermore if the diagnostic grid to calculate MFC is not representative, such as with a mesh that is much finer than the observational data and not properly smoothed, MFC features in turn will be smoothed out or distorted.

There is also evidence that the contributions of convergence to convective initiation may be rather significant and are not represented well by MFC expressions. Forecasters may be better advised to simply examine the pure mass convergence fields rather than rely on MFC fields. Nevertheless, this does not downplay the importance of moisture; not only does it contribute to more parcel buoyancy but it has also been shown by numerical modelling (Richardson 1999) that storms preferentially propagate toward areas of higher moisture, i.e. higher dewpoints or theta-e.

2.4. "POOLING". Moisture pooling is a metaphor for a "blob" of high dewpoints which seems to form at a particular place on the weather map, often along or just north of a warm front during the summer. This is not an actual physical process because water vapor is part of the air itself and cannot be increased just by converging.

The explanation for moisture pooling is that in convergent areas the boundary layer deepens, retarding moisture loss through vertical mixing. Over time, a localized dewpoint maximum forms at this location. No new moisture is added. However dewpoints may be influenced by outside sources, such as irrigated fields.

2.5. MESOSCALE STRUCTURE. Field experiments and general observation indicate that the tropical air mass is not homogenous in a pre-storm environment. The depth of the moist air mass often has an undulating quality, shaped by the vertical circulations within the air mass. This tends to produce favored areas of initiation. Though exactly how these processes correlate to convection is not well understood, it underscores the influence that cloud-scale circulations have on the forecast.

3. Pressure

Barometric pressure is intimately tied to the movement of air in the atmosphere, and reflects the thermal properties of the atmosphere. While the actual pressure reading has no real significance in forecasting, trends over time and patterns across a horizontal area can reveal a wealth of detail about a storm environment.

3.1. ALTIMETER SETTING (ALSTG) (QNH). This is one of two key pressure expressions used in weather analysis. The pressure is adjusted to sea level using a simple mathematical formula. It is expressed in inches of mercury and is available at most stations. It should be used when tracking small-scale disturbances, since it is a pure measure of pressure, and since it is reported at nearly all stations it can provide maximum resolution for isopleth analysis. Since there are no temperature corrections in ALSTG computation, it is very sensitive to diurnal heating factors. Magor (1958) emphasized in mesoanalysis that altimeter setting provides "a more spontaneous indication of true pressure behavior in and around a meso-structure" than sea-level pressure offers.

3.2. SEA-LEVEL PRESSURE (SLP) (QFF). This pressure expression forms the basis for about 90% of surface charts found on the Internet. The pressure is adjusted to sea level using a standard atmo-

---

Cyclone types

* Mesolow: A cyclone measuring about 10 to 200 miles in diameter appearing on mesoscale charts. Sometimes called sub-synoptic lows.

* Mesocyclone: A large vortex within a convective storm measuring 1.5 to 6 miles (2 to 10 km) in diameter.

* Misocyclone: A small vortex within a convective storm measuring about 150 ft to 2 miles (0.04 to 4 km) in diameter, but generally larger than the scale of a tornado.

* Subsynoptic low: Any low smaller than synoptic scale.

The origin of mesolows

Barogram traces from a tornado that crossed northern Ohio in 1953 uncovered that the storm was associated with a low pressure area about 20 miles in diameter (Lewis and Perkins 1953). Forecasters named these cyclones mesolows (Magor 1958) to distinguish them from the microscale "tornado cyclone" (later named "mesocyclone") and the macroscale extratropical low.

sphere and using current temperature and that of 12 hours previously to try to eliminate diurnal factors. It is expressed in millibars. SLP should not be used in severe thunderstorm nowcasting, as the 12-hour temperature corrections are designed to dampen meso-scale influences. Also since some stations do not have reduction algorithms computed, a sizable fraction of stations do not report SLP and this can diminish the resolution of analyzed isopleths.

3.3. MESOLOW. Surface analysis may reveal the existence of very small-scale lows measuring on the order of many tens of miles in diameter. These may move across a storm target area and interact with an area of convection. In severe weather forecasting these are referred to as *mesolows*. The mesolow is caused by small scale processes in the atmosphere and may be supported by mesoscale processes aloft.

The northeast quadrant$_{(NH)}$ of a mesolow, because of the backed winds in this area, contains the strongest values of shear, and may cause one particular area of storms to become tornadic. Magor in 1958 observed that mesolows are "so transitory that they can only infrequently be observed on routinely prepared synoptic charts", and to a certain extent this is still true, underscoring the value of mesoanalysis.

Excellent technique for identifying areas of mesolow development is by monitoring trends of wind and pressure at individual stations and analyzing/tracking areas of mesoscale pressure falls (isallobaric analysis). Unlike the conventionally-prescribed 3-hour sea-level pressure change, severe weather forecasters should use 1-hour altimeter setting change. The small-scale lows revealed by detailed analysis are barely resolvable, occurring on a scale of hours, 20 to 200 miles in size, and often with pressure changes of as little as 0.03 to 0.06 in/hr. Moller 1979 documented "pressure fall waves" that were detectable 1 to 2 hours before the 1979 Wichita Falls outbreak and were useful in identifying mesolow centers.

3.4. MESOHIGH. The outflow area of a thunderstorm produces a localized high pressure area underneath the storm, sometimes called a "bubble high" due to its three-dimensional bubble shape and its tendency to expand. This bubble high contains rain-cooled air. It may expand to an area larger than the storm's areal coverage, resulting in what is known as a *mesohigh*. A large MCS may produce a mesohigh much larger than the storm's areal coverage, and it can continue to exist for days after storms have dissipated. The edge of the mesohigh is demarcated by an outflow boundary.

4. Fronts

In severe weather forecasting, boundaries are important. They represent an axis where some sort of convergence or lift is taking place. They are significant for two key reasons: they provide a forcing mechanism for initiation and they provide a source of vertical vorticity for non-tornadic misocyclones and horizontal vorticity for tornadic mesocyclones.

A front is a transition zone separating two air masses of differing density. Thus a temperature gradient is the primary indicator of a front. Wind shift lines, troughs, precipitation bands, and other features are only secondary indicators, and if they exist alone then by definition the feature is not a front. Since temperature varies depending on station altitude, an analysis of potential temperature (theta) is often used to normalize the temperature field across a forecast area and help locate the front.

4.1. WARM FRONT. Since the warm front is often perpendicular to the tropospheric winds, storms tend to depart the warm front boundary and enter the cold air mass, where they ingest colder air and weaken. The width of the severe weather window along the warm front is inversely proportional to the strength and depth of the cold air and can range from just a few miles to 100 miles or more. When the mean tropospheric winds are parallel to the warm front, there is a greater likelihood for storms to be persistent, though the precipitation cascade can interfere with other storms forming along the boundary. Some of the strongest warm front outbreaks are associated with mean upper tropospheric winds that are neither perpendicular nor parallel to the warm front axis, but somewhat diagonal, which deposits the precipitation on the cool side of the front but allows for deviant propagation of the updraft parallel with the boundary.

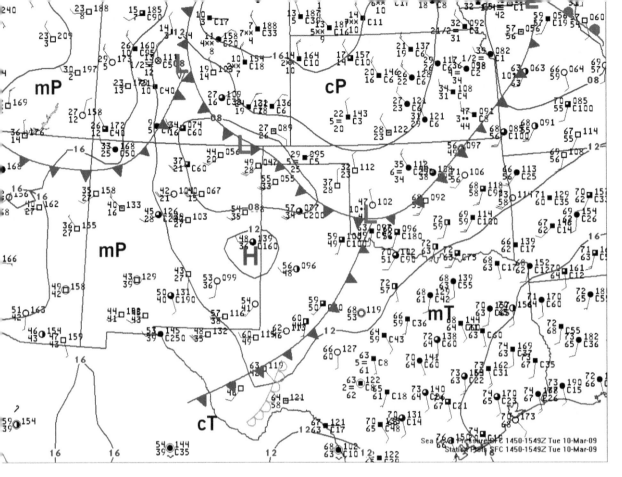

Figure 10-4. Two types of Great Plains fronts are illustrated on this map from 10 March 2009. The Canadian polar front above is found in this example from Oklahoma to the Texas Panhandle and ushers in frigid, shallow continental polar (cP) air. Further west are two Pacific fronts crossing the mountains, bringing deep, mild Pacific air, noted with the maritime polar (mP) designation. *(Tim Vasquez / graphics Digital Atmosphere <weathergraphics.com>)*

4.2. COLD FRONT. A well-defined cold front acts similar to a very strong outflow boundary, producing very intense linear forcing and linear storm modes. This is especially true if tropospheric flow is parallel to the boundary, which causes precipitation cascades to fall along or behind the front, reinforcing it.

Though the cold front is itself in always located in a pressure trough, pre-frontal troughs can precede cold fronts and be an important source for initiation. During the mid-20th century, many squall lines were classified as being "pre-frontal", that is, occurring up to 100 miles or more ahead of the cold front. It is believed that gravity waves initiated off the cold dome behind the front can also break the cap in the ascending phase further downstream when the upper flow is has a perpendicular component to the front and the western portions of the cold air mass are blocked by the Rockies or are simply on the westward-sloped terrain of the Great Plains (Ralph et al 1999).

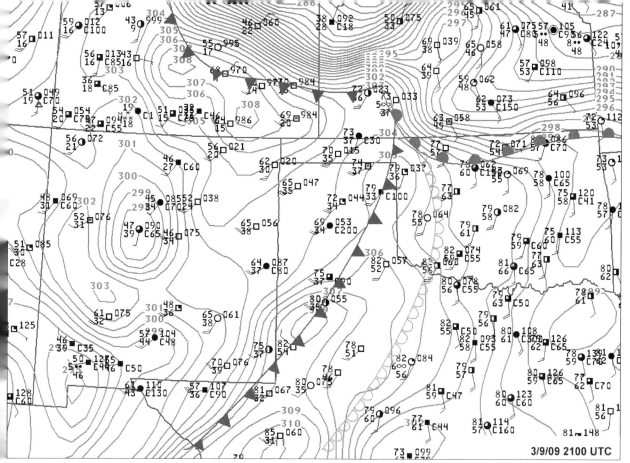

Figure 10-5. Weather map showing a cold front emerging from the Rocky Mountains. A dual cold front - dryline pair is quite common, with the cold front eventually overtaking the dryline. Note that potential temperature (theta) isopleths help to isolate frontal positions quite effectively. (Tim Vasquez / graphics Digital Atmosphere <weathergraphics.com>)

4.3. DOWNSLOPE EFFECTS. In the Great Plains, analysts frequently have to deal with cold fronts located within the Rocky Mountains, where terrain and varying elevation can disrupt thermal fields, wind shifts, and other gradients. Furthermore, when the cold front emerges and begins its descent into lower elevations, the cold air mass advecting eastward moves downslope and warms adiabatically. This can reduce temperature contrasts along the front and make it very difficult to find.

One tool for locating the front is potential temperature. This normalizes all temperature readings to a common level (1000 mb). But better yet one of the most effective tool for monitoring these troublesome fronts is time continuity: using several hours of history for surface maps and station reports. A noticeable wind shift at Albuquerque, accompanied by a slight temperature fall, suggests one example of a subtle frontal passage that might be missed by a standard map analysis.

5. Other boundaries

One key reason why boundaries are important is that any temperature gradient, even if the cause is not a clear contrast of air masses, alters the low-level wind field producing a zone of enhanced horizontal vorticity. This vorticity zone can be converted to streamwise vorticity and ingested by storms, producing supercellular modes. Even after the original temperature contrasts have disappeared, the wind field can remain in an altered state with the horizontal vorticity still be present.

5.1. NEW OUTFLOW. The thunderstorm downdraft produces significant amounts of outflow with the leading edge marked by a gust front or outflow boundary. This can be sufficient to regenerate new cells and is how many multicell clusters propagate. From a tornado forecasting standpoint, fresh outflow such as that along a rain-cooled forward flank downdraft edge does not typically produce the parcel residence times required to develop streamwise vorticity that leads to a mesocyclone.

5.2. SHADOW BOUNDARIES. The shadow produced by a dense anvil may produce a localized area of cooling, resulting in a weak, localized frontal boundary where it contrasts with strongly-heated air surrounding the storm. Research has begun focusing on the importance of this mechanism.

5.3. STAGNANT OUTFLOW. Outflow from thunderstorms may survive for hours or even days after thunderstorms dissipate, and may take on frontal characteristics. The boundaries delineating these outflow pools are often visible on sensitive radar reflectivity imagery as the day progresses, and during the midday hours on high-resolution visible imagery as cumulus fields form.

5.4. BOUNDARY INTERACTION. A supercell that crosses a boundary or a gust front, or interacts with it, is a candidate for tornadogenesis. Storm mergers can lead to gust front interaction. On the other hand, a gust front which spreads away from the storm is not a candidate for tornadogenesis.

DIAGNOSIS 217

One recent study has found that at least in the case of landspouts, the existence of one stationary boundary is associated with an increased incidence of landspouts. Two moving boundaries, by contrast, seems to reduce the landspout potential.

6. Dryline

The dryline is common in the southern Great Plains and marks the division between moist tropical air from the Gulf of Mexico and dry continental air originating from the plateau regions of the southwest U.S. and northern Mexico. The dryline also occurs in other parts of the world where warm tropical air interacts with dry continental air, but this text will focus exclusively on the Great Plains dryline.

When cyclogenesis and lee side troughing occurs in the Rocky Mountain region, the convergence focuses air mass contrasts, and so the demarcation between dry and tropical air strengthens. Since the tropical air usually has only a limited depth, on the order of 1 to 2 km in the spring, the dry air mass often overlies the tropical air mass. This not only caps it but forms a layer known as the elevated mixed layer (EML). Since the elevated layer has warmer characteristics, it forms an inversion.

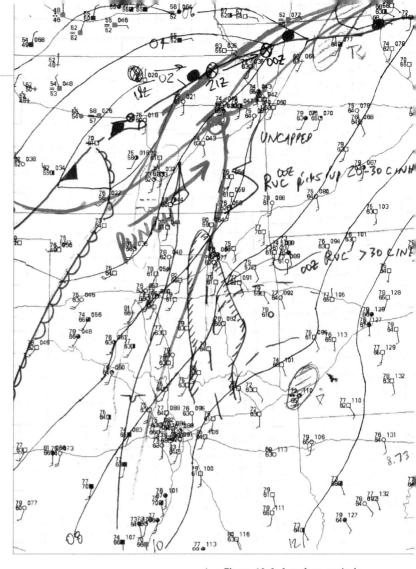

Figure 10-6. A surface analysis does not need to be neat but should be very comprehensive and serve as a canvas for thought. A variety of markings can be seen here which helped form the diagnosis of surface conditions. *(Tim Vasquez / graphics Digital Atmosphere <weathergraphics.com>)*

6.1. ANALYSIS GUIDELINES. The dryline should be drawn along the moist side of the moisture gradient that divides the tropical from the dry air mass. When no clear dryline location can be found, the 55-degree isodrosotherm may be used as a *first guess* (Schaefer 1986), but this rarely correlates with the actual dryline location and must always be refined. Multiple drylines may be drawn if it is unclear which one is significant.

The dryline is always located by using gradients of specific humidity, i.e. dewpoint or mixing ratio. Relative humidity and equivalent potential temperature should be avoided since these have temperature dependencies. The dryline should not be located using temperature gradients and wind shifts, and furthermore it does not necessarily lie in a pressure trough (Koch and McCarthy 1982).

Visible satellite and radar imagery should be used to refine the position of the dryline. Visible satellite imagery will be useless in situations where considerable cirrus debris exists, but radar often shows a distinct fine line if the dryline is strong enough or close enough to the radar site.

If a Pacific front is in the region, it may be difficult to differentiate the dry sector airmass from the Pacific polar air mass. Cool daytime temperatures, marginally dry dewpoints, and considerable low or mid-level overcast all are indicative of a polar air mass. Potential temperature gradients within the air mass can help isolate the location of a difficult-to-find Pacific front.

6.2. ACTIVE DRYLINE. Dryline storms are most common in Texas, eastern New Mexico, and western Oklahoma. Storm initiation is often enhanced where the dryline is superimposed with the Texas Caprock east of Amarillo and Lubbock, a source of enhanced lift. A significant westerly component in the low level winds will thin out the moisture, allowing the dryline to mix out moisture more rapidly and causing it to move eastward. This process is sometimes referred to as scouring, and is most common when significant pressure rises occur along the Rocky Mountains, such as with the departure of an extratropical cyclone.

Very strong, fast-moving drylines often have a significant westerly component in both the moist and dry sector. This is associated with subsidence due to air moving to lower elevations, but can produce strong heating in the dry sector.

6.3. RETREATING DRYLINE. A dryline typically retreats westward when the sun goes down and vertical mixing stops. In other cases, strengthening surface moisture advection is the dominant factor, a situation which may occur with the approach of an upper-level disturbance. The winds usually have an increasing easterly component, and on the Great Plains this is usually is in an upslope direction, which adds to lift and instability. Storms can even develop and intensify, such as the 11 May 1970 event in Lubbock where tornadoes touched down after dark.

Later at night, radiational cooling in the dry, clear air mass is strong and temperatures can fall well below that in the tropical air mass. This makes the early morning analysis challenging and masks the true location of Pacific cold fronts that may be nosing into the dry air mass.

6.4. BULGES. Dryline bulges are identified as portions of the dryline which push strongly into the tropical air and are often associated with highly ageostrophic flow (Tegtmeier 1974). It has been proposed that they are associated with the development of mesolows (Livingston 1983), which provide localized sources of convergence and strong low-level shear for severe storms. One theory for the bulge and enhanced convergence downstream from the bulge along the dryline (Hane and Richter 2004) is locally enhanced vertical mixing, which could occur from very strong heating over very dry or vegetation-sparse terrain.

Recent work has identified the existence of misocyclones (on the scale of single kilometers) along the dryline, which are too small to be resolved with the surface network but can sometimes be detected in radar fine lines. These misocyclones are believed to be influential in shaping the moisture field along the dryline.

6.5. INITIATION MECHANISMS. The dryline serves as a focus for severe thunderstorms primarily because it is a zone of intense convergence separating dry westerly flow from moist southeasterly flow. This convergence allows moisture to deepen and increases the chance of moist parcels ascending to their LFC. Also the dryline contains a baroclinic solenoidal circulation, much like a sea breeze, due to the difference in virtual potential temperature. It results in a zone of lift along the dryline.

7. Subjective diagnosis

While severe weather forecasting prescribes the ingredient-based method of diagnosis, there are subjective elements that are also useful in the forecast process.

7.1. PATTERNS. Humans instinctually see patterns in almost everything. Carl Sagan said "the pattern recognition machinery in our brains is so efficient that we assemble disconnected patches of light and dark and unconsciously try to see a face." During the early 20th century, computers did not yet exist, so forecast efforts heavily emphasized pattern recognition. For severe weather forecasters this culminated with the system developed by Col. Robert Miller in the 1960s which outlined severe weather outcomes by pattern. The technique became marginalized by the 1970s with the advent of objective methods, ingredient-based forecasting, and a greatly improved understanding of convection. Even with these great advances, pattern recognition is still a deeply ingrained part of the forecast process, even if only intuitively. While there is no pattern-based forecast system that is appropriate for severe weather forecasting, we do provide a framework for pattern-based diagnosis in the next major section, "Surface Target Zones".

7.2. CONTINUITY. The simplest form of subjective analysis is continuity. According to continuity, if no significant changes occur with the weather pattern, the same weather that happened yesterday will occur again today. More significant is the fact that minor changes can be accounted for by biasing the result: if upper-level winds weaken, it is safe to assume that compared to yesterday storms will be more slow-moving, more HPish (if supercells develop), and upper-level disturbances will be weaker. It's also possible that heights might be a little higher so the cap might be stronger. A noteworthy example of continuity is the May 8, 2003 tornado event in Oklahoma City, which produced a tornado across Moore and Tinker Air Force Base. The very next day, on May 9, the pattern did not change significantly and more tornadoes struck Oklahoma City's north side.

Miller's patterns

These patterns were defined in the 1960s by Col. Robert C. Miller to aid in forecasting. The classifications are not used anymore but are referred to by many late 20th century journal articles on severe weather, and still serve as useful archetypes.

* **Type A**. Southwesterly jet, dry mid-level intrusion, associated with classic Great Plains tornado outbreaks.

* **Type B**. Deep upper level low to the west, progressive cold front. A squall line is likely.

* **Type C**. Quasistationary front, dry mid-level air, and westerly winds aloft. A few tornadoes typically occur.

* **Type D**. Closed low, cold-core situation. Hail is the main threat.

* **Type E**. Quasistationary front, strong cyclogenesis. A squall line is likely.

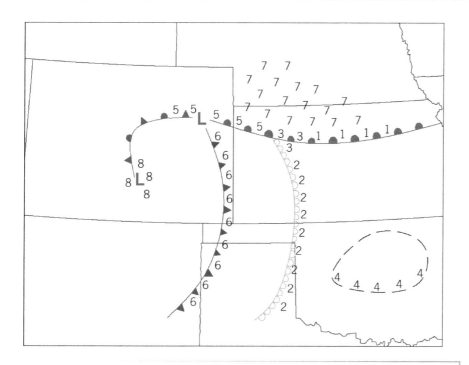

Figure 10-7. Eight-zone master reference for surface features in severe thunderstorm initiation: (1) warm front; (2) dryline; (3) triple point; (4) outflow boundary; (5) upslope flow; (6) cold front; and (7) isentropic lift. In many cases the dryline does not exist or is overrun, in which case the triple point (area 3) moves to the cold front - warm front intersection and the remainder of the warm front comprises a warm front situation (area 1). See the text for details. *(Tim Vasquez)*

7.3. CLIMATOLOGY. Climatology is considered a baseline for what type of weather is most likely to occur. One example of a climatology forecasting technique is Tim Marshall's rule: "When it's May, you chase." This implies that the risk of thunderstorms is greatly in a chaser's favor in late spring. May in the Great Plains provides the greatest overlap between the high equivalent potential temperatures of the warm season and the destabilizing quasi-geostrophic disturbances of the cool season. The "day before the big day" is often cited as a peak for chaseable activity; part of the reason for this is storm seeding and interactions are fewer, and tropospheric flow is slower, making storms more chaseable. However climatology can also be a trap, lulling a forecaster into believing that a severe storm in December, for example, is impossible since it has never occurred.

8. Surface target zones

The idea of surface target zones uses eight primary zones (Figure 10-7) commonly found in central United States extratropical cyclones. This concept provides forecasters, especially new ones, with a useful starting point for diagnosis. The outcomes indicated

Go to the source

A consistent way to judge the quality of a model forecast, and to develop a basic expectation on your own, is to consider the source regions for the basic ingredients relevant to forecasting severe storms. Lapse rates over Rockies, surface trajectories across the Gulf and Caribbean, remnant outflow boundaries from morning convection - all of this information can be gleaned from observations well before you look at a single model graphic.

RICH THOMPSON, 2009
personal communication

The devil is in the details

The 3 May 1999 tornado outbreak in Oklahoma has proven notoriously difficult for numerical models to reproduce even years after the fact. However, real-time diagnosis of satellite imagery and profiler wind plots helped focus attention on the area of southwest Oklahoma where supercell formation was probable. The same can be said for the Super Tuesday outbreak on 5 February 2008. Despite the availability of models that can resolve some supercell structures, the timing and location of storm formation was uncertain even as of noon the day of the event!

It turns out that a combination of surface analysis, visible satellite imagery, special soundings, and pattern recognition helped identify the likely corridors for supercell development by early afternoon. In both cases, the initial storm development proved critical to outcome, and an accurate forecast would have been hard to come by with model output alone.

RICH THOMPSON, 2009
personal communication

---

here are only typical ones. For example if the air mass is poorly capped and unstable, storms may develop along all of the boundaries and eventually consolidate into a large MCS.

8.1. WARM FRONT (ZONE 1). The warm front is perhaps the most prominent boundary encountered by a moist tropical air mass, and for this reason it is a favored area for convective initiation. If storm motion is significantly perpendicular to the warm front, storms will have a tendency to depart poleward into the cold air. Therefore storms along the warm front may have a limited window in which to produce tornadic activity.

8.2. DRYLINE (ZONE 2). The dryline is well-known for its role in Great Plains chasing, as in a relatively quiescent weather pattern it represents the demarcation line between moist tropical air and approaching upper-level weather systems. In many instances it is not the ideal location for severe weather. Since the dryline is found in the southern quadrant of the synoptic-scale low (at least in the United States), winds along the dryline tend to be southerly and in many instances the enhanced backing needed to sustain severe, rotating storms is not found here but closer to the east and northeast quadrant of the large-scale low, i.e. near the warm front and triple point areas. Still, northern segments of the dryline, segments associated with mesolows and bulges, and intersections with outflow boundaries are quite capable of producing significant severe weather outbreaks.

8.3. TRIPLE POINT (ZONE 3). Rather than defining a triple point by specific boundary types, we define the "triple point" as the *cusp of the tropical air mass*. Since the primary east-west boundary is normally baroclinic, i.e. a warm front or outflow boundary, there tends to be cyclogenesis in this area with the approach of upper-level dynamics and a rich field of streamwise vorticity. Winds in this area have a tendency to back, which further increases shear profiles. On the Great Plains, the northwest cusp of tropical air is often in high terrain, and this combined with easterly flow introduces an upslope nature to targets in this region, which further increases destabilization.

8.4. OUTFLOW BOUNDARY (ZONE 4). Stagnant outflow boundaries in the moist sector, when they occur, can be a significant source for convective initiation. *Very weak boundaries and convergence areas often produce thunderstorms in the moist sector*, so the burden is on the forecaster to identify all potential development areas. The zone where moist tropical air is impinging on the boundary is the most likely area to produce initiation and severe weather. Intersections of outflow boundaries and other boundaries with storms, including MCS lines and bow echoes, are correlated strongly with enhanced severe weather, especially when the boundary axis is parallel to storm movement.

8.5. UPSLOPE FLOW (ZONE 5). Here, upslope does not refer to flow up mountains, but rather up broadly ascending terrain. In the Great Plains, terrain ascends from east to west. Therefore in synoptic-scale easterly flow (i.e. wind from the east) air is forced to ascend to higher elevations. This is essentially a forced ascent mechanism. It increases the relative humidity of the parcel and produces lower LCLs, a favorable ingredient for rotating and tornadic storms. Storms often develop where the highest theta-e values (moisture, and to a lesser degree, temperature), work westward onto higher terrain.

8.6. COLD FRONT (ZONE 6). Cold fronts generally provide a linear wall of forcing and are very efficient at initiating convection. On the same token, they are detrimental to severe weather because cold fronts bring cold air which readily undercuts updrafts, producing outflow-dominant storms. Cold fronts can be associated with damaging or tornadic supercells, but this generally occurs only with slow-moving fronts and cold fronts with weak temperature contrasts. Cold fronts in the Great Plains usually emerge from the Rockies and are preceded by a dryline. The cold front eventually catches up to the dryline and displaces it aloft. When this happens, forecasters must be vigilant as convective modes can change rapidly.

8.7. ISENTROPIC LIFT (ZONE 7). The isentropic lift pattern occurs generally north of a warm front above the shallow cool air. It has an elevated inflow layer, which reduces the chance of tornadic

---

Conceptual models

I believe that imposing conceptual models [such as frontal notation] on the data, rather than adapting conceptual models to fit the data, is a dangerous and unscientific thing to do. We shouldn't be seeking to confirm our cherished conceptual models in operational practice, or in our scientific investigations. Rather, we should be seeking to understand the structure and evolution of the atmosphere, at least insofar as it's revealed by our data.

CHARLES DOSWELL III, 2006
Conceptual Model Concerns

activity, and low-lying stratus layers frequently obscure the cloud base. These two reasons reduce its interest to chasers. Nevertheless, substantial lapse rates above the shallow cold dome can still contribute to these storms being prolific producers of hail and wind.

8.8. COLD CORE LOW (ZONE 8). This corresponds to the Miller Type D pattern (Miller 1972). Most upper-level cold pools are co-located with closed upper-level lows, which in turn often originate from occlusion of extratropical frontal systems, placing them some distance within the cold sector or the dry slot. Because this area is not linked well with the warm sector, forecasters may underestimate the instability and be caught unaware by severe weather development (Carr and Millard 1985, Goetsch 1988). Some of this development may occur from elevated moist layers. Minisupercells are not uncommon and are favored if the low-level flow is strongly backed with height. All cold-core convection is strongly diurnal and almost always dies by sunset.

It should be noted due to the small scale of the upper-level low and the great change in quantities over small distances, diagnostic quantities from analysis and numerical models may especially troublesome around cold-core lows, especially where fields are excessively smoothed or not well correlated. Composite parameters such as the STP and EHI do not work well with cold-core events (Davies 2006).

9. Convective modes

Once we have determined where storms might form and what the likelihood of initiation is, we can consider the convective mode in full detail. This boils down to not just instability and shear, but also the relation of storm structure to boundaries.

9.1. INSTABILITY AND SHEAR. The contributions of instability and shear to a developing storm are covered in detail in the chapter on Stability & Shear. In short, strong instability and strong storm-relative helicity favor severe or rotating storms.

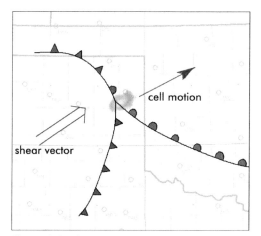 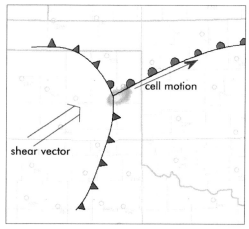

Figure 10-8a. Shear vector perpendicular to thermal boundary. Cells which form on the boundary have only a short window of severe activity before becoming elevated, but seeding of cells is reduced.

Figure 10-8b. Shear vector parallel to thermal boundary. Longer lived cells are possible since storms preserve their access to surface-based moisture, but seeding risks are increased. *(Tim Vasquez)*

9.2. ANVIL-LEVEL WIND VS. MOTION. If the *storm motion is parallel to anvil-level flow*, the updraft will tend to ingest parcels that have been within the baroclinic zone along the anvil shadow. Enhanced tornado potential is possible. This of course requires strong solar heating, an inflow trajectory relative to the anvil edge. If the *storm motion is perpendicular to anvil-level flow*, the updraft is less likely to move toward the anvil-edge baroclinic zone. This may reduce the chance of tornadogenesis.

9.3. SHEAR VECTOR VS. BOUNDARIES. The orientation of the deep-layer shear vector relative to surface boundaries has a major effect on the type of convective mode that develops (Bluestein and Weisman 2000). If the *deep shear vector is parallel to boundaries*, storms will tend to interact, particularly left-movers, and form into lines. Supercells are possible at the equatorward end of the line. Chasers refer to these end storms as "tail end charlie". Right-movers may move far ahead of the line. If the *deep shear vector is diagonal to boundaries*, discrete supercells will form; little interaction will occur. With this configuration, storms are more likely to remain isolated. If the *deep shear vector is perpendicular to boundaries*, col-

Figure 10-6. The relationship of weather regimes to surface low position was recognized as far back as 1874 in the British publication "Instructions for Meteorological Telegraphy".

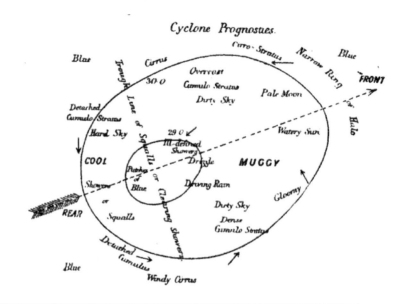

liding storm splits may occur, with right-movers and left-movers merging.

9.4. MOVEMENT VS. BOUNDARIES. If the *storm motion vector is perpendicular to the thermal boundary*, pointing into cold air, the storm will have a tendency to form along the boundary and then immediately begin moving into the cold air. This provides a window of opportunity for severe weather which is inversely proportional to the depth and intensity of the cold air. If the *storm motion vector is parallel to the thermal boundary*, the storm will have a tendency to form along the boundary and move along it.

9.5. ANVIL WINDS VS. BOUNDARIES. If the *anvil wind vector is perpendicular to a boundary*, precipitation cascades fall away from the boundary, largely failing to seed other storms on the boundary and interfere with their development. If the vector points into the cold pool then the cold pool may be augmented and generate linear storms. If the *anvil wind vector* is parallel to a boundary, precipitation cascades fall along the boundary where other storms are forming, producing numerous cells and possibly linear configurations.

10. Special characteristics

10.1. NORTHEAST QUADRANT OF A SURFACE LOW. The northeastern$_{(NH)}$ quadrant of a surface low, or southeast$_{(SH)}$, contains easterly winds that provide favorable storm-relative helicity. This frame of reference assumes southwesterly upper-level flow, so in the presence of northwesterly upper-level flow, the southeast quadrant of a surface low is more favorable. This technique depends on the existence of favorable thermodynamics in that area, so it is rarely applicable to synoptic scale lows where cold air is present but may be highly relevant to small-scale mesolows where warm, moist air is in abundance.

10.2. NORTHWEST FLOW. A well-organized complex of storms in northwesterly flow, or downstream from an upper-level ridge, is always suspect for intensification since the steering flow is usually directly into the supply of moisture. In the summer months in Canada and the northern plains, northwest flow tends to be associated with derecho events.

10.3. UPPER LEVEL PATTERNS. Evaluation of the upper-level charts is sometimes referred to as evaluation of the meteorological "dynamics". Though thermodynamics and instability can be inferred from upper-level charts, the colloquialism "dynamics" refers to the various processes associated with large-scale weather systems, better defined as quasi-geostrophic forcing. When it produces upper-level lift, the motion is usually on the order of only centimeters per second, the sole effect being to gradually destabilize the atmosphere and allow for surface pressure falls. This is by no means an actual trigger but a mechanism that improves the chance for thunderstorms to develop.

11. Localized phenomena

In mountainous regions, the mountain peaks themselves are often a focus for orographic lift, forcing the ascent of moist air on a certain flank and generating thunderstorms later in the day. On the Great Plains, or on any expansive terrain for that matter, there are

---

Northwest flow

Forecasters seem to have poor skills at diagnosing the hodograph and not being able to conceptualize actual storm motion in the field, given the hodograph that exists. I still see forecasters positioning themselves on the south or southwest flank of a storm, when the hodograph suggests that the whole "hook" structure would be rotated so that the wall cloud would be on the northwest side of the radar echo (for northwest flow types).

JOHN MONTEVERDI, 2009
personal communication

Finley's tornado criteria

John P. Finley in 1888 established the very first tornado forecasting rules:

1. The presence of a well-defined area of low pressure; marked barometric gradient not necessary.
2. Slow progressive movement of the low in order to increase the flow northward of heat and moisture in the southeast quadrant [of the low].
3. A troughlike low trending north and south, or northeast and southwest.
4. The descent [i.e. movement southward] of a well-marked anticyclone in rear of the low.
5. High temperature gradients.
6. The increasing wind velocities of the southeast, southwest, and northwest quadrants of the low.
7. The northward curve of the isotherms in the southeast quadrant and the eastern portion of the southwest quadrant of the low.
8. The southward curve of the isobars in the northwest quadrant and the northern portion of the southwest quadrant.
9. The high temperature gradient between apices of the opposing curves of temperature.
10. The increasing and uniformly high humidity of the southeast quadrant.

Finley also emphasized climatology, the affinity for the southeast quadrant of a low several hundred miles southeast of its center, and the eastward curve of the wind shift line corresponding to the cold front.

---

also a number of mesoscale discontinuities that have an effect on the weather. Some not mentioned in this book include certain river basins, like where the Canadian River cuts through the Texas Caprock. This is suspected to produce a local enhancement of tornado production, perhaps in response to mesoscale terrain-induced circulations. Prominent mesoscale and synoptic features include the ones below:

11.1. LLANO ESTACADO AND TEXAS CAPROCK. The Llano Estacado (Staked Plain) is a large plateau extending from west Texas through the Texas Panhandle, tapering northward into Kansas. Though the terrain itself is washboard-flat, the eastern slopes are marked by rugged, rapidly rising terrain that increases dramatically from 1500 to 3000 ft MSL. This slope is known as the Texas Caprock, which supports a north-south axis of enhanced lift extending from the eastern Texas Panhandle southward to Snyder, Texas. Early studies during the late 1970s found thunderstorm maxima along the Caprock and proposed that enhanced convergence and orographic lift were important mechanisms along its length. The plateau itself has a dense farming and ranching economy, and also has an western slope in New Mexico known as the Mescalero Escarpment.

11.2. PALMER DIVIDE. Located in Colorado, the Palmer Divide is a large ridge about 1000-2500 ft higher than the surrounding terrain occupying a triangle between Denver, Colorado Springs, and Limon. It represents a source of orographic lift, especially when moist low-level flow is coming from the south or east quadrants. It is a favored area for convective development. Storms that form here have a tendency to track northeast or east toward the Limon area.

11.3. DENVER CONVERGENCE - VORTICITY ZONE (DCVZ). The DCVZ forms in northeast Colorado when low-level flow is out of the southeast. It is caused by the Palmer Divide north of Colorado Springs and the Front Range of the Rocky Mountains. This produces a vorticity maximum centered on Denver. An axis of convergence usually extends northeast across the Colorado High Plains and may contain small-scale cyclones. The DCVZ is a noteworthy producer of non-supercell tornadoes on marginal severe weather days. However it is not a prolific tornado producer since the higher shears that are associated with tornadic storms usually cause these storms to initiate in the DCVZ and move eastward out of the local area.

11.4. RATON RIDGE CONVERGENCE - VORTICITY ZONE. There is some evidence that the Raton Mesa is a favored area for storm development. The convergence zone is somewhere in the region northeast of Trinidad, Colorado, analogous to the DCVZ in that it is associated with an east-west ridge (the Raton Ridge) and a north-south ridge to the west (the Sangre de Cristos). Though radar appears to confirm the idea, the weather observation network is too coarse in this area to study it further. Its influence on storm days sometimes extends from Trinidad and Raton east toward Springfield and the Kansas border.

11.5. SERRANIAS DEL BURRO. The Serranias del Burro is a mountain range located in Mexico halfway between Del Rio, Texas and the Texas Big Bend. Its eastern slopes, rising to 5500 to 7000 ft MSL, serve as a potent source of upslope lift in an area that is often strongly capped. It is often a prime initiation area for supercell thunderstorms (see Edwards 2006). These storms, if they move eastward, track into populated areas of Coahuila and into the Del Rio region.

11.6. NORTH AMERICAN MONSOON (NAM). Also known as the "southwest monsoon", this pattern describes the runup in thunderstorm frequency in Arizona and adjoining areas of neighboring states in the late summer. It not only provides much of the important annual precipitation and influences the wildfire threat, but presents flash flood risks and for some chasers provides photogenic lightning opportunities. In Arizona, the monsoon generally arrives in late June or in July and ends in September.

The Gulf of Mexico has traditionally been considered the moisture source for NAM cases, and indeed this is true for much of New Mexico, which sees the arrival of Texas moisture in June and July with the retreat of the dryline westward. But in Arizona, evidences strongly suggests that the Gulf of California (GC) is a key source of moisture, where sea surface temperatures may average up to 85°F.

Figure 10-7. A Serranias del Burro storm, which has moved off the mountains and into the Del Rio area. *(NOAA)*

Thunderstorm activity is tied strongly with the arrival of surges of moisture from the GC area (Bright and McCollum 1998). Other important moisture sources are evapotranspiration in the Mexican interior, and especially in the eastern regions, moisture from the Gulf of New Mexico. Thunderstorms almost always initiate orographically but as they evolve they tend to "detach" from the mountains and move with the prevailing flow. Thunderstorm activity may rapidly increase in coverage from one day to the next, known as a burst, then temporarily recede in what is known as a break.

One of the most significant problems with the NAM model is the lack of observed data, as there are usually no radiosonde sites operational in the Gulf of California and only a few surface sites. Forecasting is heavily dependent on satellite data and model output, however it must be kept in mind that without a good observational network in place, there is really no way for the model to know what is really going on in northwest Mexico, regardless of how state-of-the-art the model is.

11.7. LOCALLY MOIST AREAS. Localized (mesoscale-beta and gamma) areas of recent rainfall, lush woodland, and irrigated cropland create localized enhancement of boundary layer moisture compared to surrounding areas. The extra moisture is added to the lower portions of the boundary layer, which becomes mixed throughout the boundary layer itself. The result is lowered LCL heights and increased CAPE. For the most part, the enhancements are not believed to be very significant but the topic is still an area of ongoing study. Irrigated crop fields in the corn belt during late summer are a particularly noteworthy cause of local spikes in dewpoint temperature.

11.8 LOCALLY DRY AREAS. Locally (mesoscale-beta and gamma) dry areas, such as patches of sandy or undeveloped terrain, reach higher temperatures much faster than surrounding areas. It has been shown in areas behind a dryline that this can enhance the downward transport of momentum, producing a tendency toward surface divergence near these arid locations which augments dryline bulges, and which in turn can enhance convergence further to the east. Very dry, hot areas can sometimes be identified on drought monitor and soil moisture websites, but year-round warm, dry spots also exist such as the sandy plain of the southeast corner of New Mexico.

12. Other phenomena

12.1. HAZE. Haze diminishes visibility of storms and may have an effect on spotter operations. It is caused by extremely small particles of sulfur dioxide and pollen. Contrary to popular belief, the primary source of haze (not counting the so-called "Mexican smoke") is not the tropics but the Ohio Valley region and the industrial regions of Asia and Europe. The haze problem worsens

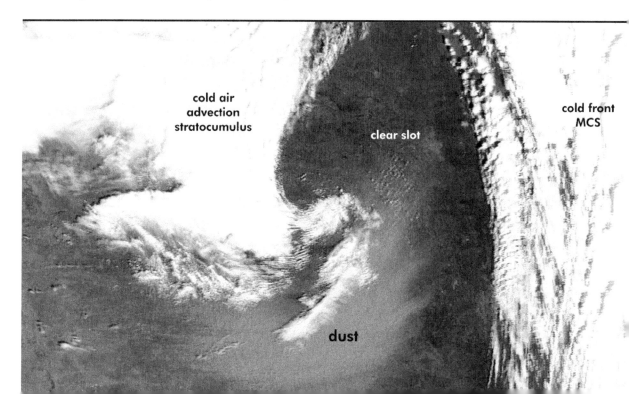

Figure 10-8. Dust accompanies many severe weather outbreaks in the Great Plains. On monochrome visible imagery it appears as a dull, streaked smudge as shown here in the lower center. Dust is usually lofted skyward in the windy, unstable dry sector (clear slot). It may also occur behind dry cold fronts, as was often the case during the 1930s dust bowl. This example occurred 24 February 2007, with the dust occupying a triangle between Wichita Falls, Abilene, and Fort Worth. About two hours later DFW Airport was shut down. *(NASA/GSFC)*

The new generation

The high-resolution models do not have much in the way of verification stats available. We at SPC have noticed a tendency for the WRF-NMM to be more aggressive with convective initiation compared to the WRF-ARW (NSSL run). These models are also quite efficient at generating atmospheric responses to modeled convection, with small bullseyes of SRH accompanying storm clusters, for example.

Still, it's not clear that these characteristics are always helpful in forecast mode. The NMM showed little precip in the warm sector from 5 Feb 2008, while the ARW was way too early on the timing and too far west. I also recall a case (5/23/07?) in the TX Panhandle where the NMM forecasted the front to sag southward all day and completely removed the warm sector from the stronger flow aloft.

It turns out that was nothing more than a model error that we, fortunately, discounted in the outlook process.

RICH THOMPSON, 2009
personal communication

---

when tropospheric wind speeds across source regions are weak or advect the pollutants into a new area, and when these winds on successive days fails to move the air mass any appreciable distance. New satellite technology makes it possible to track haze problems.

12.2. SMOKE. Smoke is another problem that may interfere with storm viewing by spotters and chasers. A very common source of smoke in the United States is from southern Mexico and Central America, caused by widespread burning of crop fields which fertilizes the soil and prepares the fields for replanting. The smoke is advected by the existing low-level wind trajectories. In May 1998, Mexican haze was a significant problem in the southern United States and frequently curbed visibility to just a few miles. Deep 850 mb flow from Mexico is often associated with these smoke events.

13. Dynamic forecasts

The idea of dynamical numerical models was first conceived in the early 20th century, but was not possible until the 1940s and 1950s with the advent of electronic computers. Since then, numerical models have reached a very high level of sophistication. Still, models are only now capable of producing reasonable forecasts of thunderstorm location, especially 6 to 12 hours in advance. But no model is currently configured to predict tornadoes, hail, or high wind.

The inherent problem with dynamical models is that they are linear in nature and the atmosphere is a nonlinear system. Therefore they are more of a coarse simulation of the atmosphere. In a nonlinear system, all elements are interdependent and very small differences between the initial conditions of a forecast model and the initial conditions of the real atmosphere will rapidly grow into large differences with time. Since it is impossible to resolve all of the differences without having omniscient knowledge of every molecule in the system, there will always be some degree of error in the prediction. Model schemes seek to minimize these differences.

Since the human brain thinks in a non-linear fashion, drawing on intuition and experience, it is especially well placed to capitalize on the shortfalls of numerical models. This is occasionally referred to in meteorological literature as the "man-machine mix".

13.1. MODEL TYPES. As of 2009 the United States weather agencies operate three key numerical models: the RUC for nowcasting, the NAM (WRF) for short-range forecasting, and the GFS for medium-range forecasting. Any of these models can be used for looking at storm forecasts during the next 12 hours, but because of the coarser resolution and global domains of the longer-range models, they are not suited for mesoanalysis. However it is possible to use the synoptic-scale forecast to identify patterns and processes that influence smaller-scale features.

13.2. MODEL SHORTFALLS. Models do have a number of inherent problems, and the ones that severe weather forecasters need to be aware of are outlined here:

• *Inadequate sampling.* In spite of the ultra-high resolution touted by many models, there are not enough observations to take advantage of the improvements. In the U.S. and Canada, observation spacing averages 80 km at the surface and 500 km aloft. Some filler data comes from aircraft and satellite data. Furthermore the new WRF/NMM model does not ingest any mesonet data yet due to perceived quality problems that would throw fields off balance. While this concern is justified in principle, carefully sited systems like the Oklahoma mesonet are not ingested.

• *Inadequate parameterization.* As model resolutions become smaller, parameterization of convective processes becomes even more important. Unfortunately there is no good scheme that can capture what exactly is taking place in a storm cloud. Furthermore at these smaller scales, cloud physics rather than wind forces become important. Radiation, water phases, cellular motion, and presence of particulate matter are not well-sampled and this adds considerable uncertainty to model physics.

• *Prioritization of model skills.* The suite of numerical models run by NCEP are largely tuned to increasing skill on precipitation and temperature forecasts. There has been long-standing criticism by severe weather forecasters that verification of moisture and stability performance is chronically neglected.

• *Lack of applicability to storm-scale forecasting.* Models excel at forecasting the basic ingredients that exist in the atmosphere prior to storm development. However they are unable to interpret the

NSSFC quotes
Humility and wisdom abound in these quotes from the 1970s-era National Severe Storms Forecast Center (now SPC).

"I've seen people die on the wrong side of the jet."
— Jack Hales

"We always issue one watch too many."
— Clarence David

"It's not as clear to me as it was at the beginning of the shift."
— Leon Schirn

"If it's April, it must be Texas."
— Popular office saying

processes that occur within a storm or handle boundary interactions well. They cannot make any sort of inherent predictions about hail, tornadoes, or flash flooding.

13.3. INITIALIZATION. The analysis of most models are currently quite excellent, but it's always worthwhile to inspect the 00-hour panel against current data. Numerical models have become quite accurate at analyzing the pressure pattern, but sometimes discrepancies exist, especially with subsynoptic features. These should always be considered in the forecast if there is any doubt that they are being handled by the model.

Consistency in the upper level charts can be a good way to check model initialization. Inspect the raw 500 mb heights/plots. Model problems are indicated by unaccountable anomalies in the relationship between observed station winds, initialized analysis, and initialized vorticity. Another item to look for is cold troughs on the raw analysis with no corresponding height troughing on the model analysis.

Models have the most trouble with jet positioning in data-sparse areas such as the Pacific Coast and northwest Mexico. Water vapor imagery combined with AIREP/ACARS data can pinpoint these features, identifying any problems with model positioning.

13.4. LARGE-SCALE LIFT. Synoptic-scale vertical motion acts to destabilize the atmosphere by lifting air and causing adiabatic cooling. This results in destabilization either by lifting the upper portion of the atmosphere, causing cooler air to overlie warmer air at the surface, or causes potential instability to be realized by allowing parcels in the lower levels to release latent heat, causing warmer air to underlie cooler air.

It's still very useful for the forecaster to have the skills to diagnose vertical motion, even on forecast charts, from basic fields of wind and temperature and from simple diagnostic methods.

Vorticity analysis has long been the staple technique for assessing vertical motion, a technique which is solidly based upon quasi-geostrophic theory. It is not a reliable technique since it makes assumptions about thermal advection and vorticity change with height, but in general the rule about cyclonic vorticity advection equalling upward vertical motion does hold true. Since vorticity centers are very well defined and move rapidly with the upper-level flow, they are useful in establishing the timing of precipitation onset.

Q vector analysis is useful for determining upward motion. Unfortunately very few such charts are found on the Internet. Other techniques for subjective analysis of vertical motion are conceptual in nature, such as the four-quadrant model described elsewhere in this book.

Another method of assessing vertical motion is to simply figure out where vertical motion is occurring in the models. Such output is given on "vertical motion" panels or depictions of "omega". Here the problem is that most weather is inherently horizontal, the scale of motion in the vertical is extremely small relative to horizontal motion, and the model fields are inherently noisy especially at the start of the run. A glance at any vertical motion output usually shows a very chaotic and ambiguous depiction of where rising motion is occurring.

13.5. SMALL-SCALE LIFT. Even if the forecast model correctly lays out all of the large-scale forcing mechanisms, the forecaster is still burdened with the problem of determining where storms will develop. Given the resolution of models and the complexity of mesoscale patterns it is very difficult to determine from model output where sufficient enhanced lift, heating, and/or cap erosion will occur to allow for deep convection.

Models are fairly good about outlining the position of fronts, boundaries, drylines, and wind shift. However even if the placement on these features is confident this poses the question of where along these boundaries, or even on which ones, convection will occur.

One shortcut method is to simply examine where the model "breaks out" precipitation. The problem with this is that precipitation forecasts are a twofold layer of assumptions: assumptions of precipitation probability based upon assumptions of the state of the atmosphere. Also precipitation fields are based more upon the end results of vertical motion and destabilization rather than where these processes are actually occurring. For a forecaster, this is like steering the car based upon bumps in the road and noises from

Model skill in 2009

I think the models have gotten very good in anticipating the synoptic fields even several days in advance. But they have failed miserably in anticipating key characteristics of the shear and buoyancy environment. The NAM, even the WRF-NAM, still appears to overforecast the moisture field systematically, possibly due to some boundary layer evapotranspiration bias.

The 0-1 shear environment is key in distinguishing non-tornadic from tornadic supercells in otherwise favorable shear environments, and both the RUC and the WRF-NAM appear to have problems anticipating that.

Finally, the models will never be able to anticipate mesoscale "accidents", e.g. exact nature of outflow boundaries and how they interact favorably or unfavorably with neighboring storms.

JOHN MONTEVERDI, 2009
personal communication

The subtropical jet

Numerical models historically have had a tendency to overestimate cap erosion with a subtropical jet, presumably because the destabilization is not as strong or deep as with a polar front jet.

Forecasting in 1883

It is absolutely impossible to predict a storm for more than a few days in advance. The information can not be too widely distributed, that no one can foretell even the general character of a coming season, much less the occurrence of a particular storm in that season. It is possible that the advance of our knowledge may at some time enable us to predict the weather for many days in advance, but this is not possible at the present time.

Meteorology is yet in its infancy, and no one is yet able to anticipate the occurrence of a meteorological phenomenon for more than a few days — a week at the most. If any one will take the trouble to verify the weather predictions which in these days are so frequently made [by charlatans] he will find that about half of them are fulfilled and half fail.

All predictions of the weather to be expected a month or more in advance, whether based on the position of the planets, or of the moon, or upon the number of sunspots, or upon any supposed low of periodicity of natural phenomena, or upon any hypothesis whatever which today has its advocates, are unreliable as predictions of the time when the end of the world will come.

WILLIAM B. HAZEN, 1883
U. S. Army Chief Signal Officer
(top federal weather post, 1883)

---

grating other cars rather than looking out the window and looking at the lay of the road and where the other cars are.

Along nonfrontal features like stagnant boundaries, drylines, and wind shift lines, forecasters can look at the convergence of the wind field. A strong wind field showing a directional shift is much more likely to result in initiation than a strong wind field where the direction is uniform.

Unfortunately aside from these there are very few other methods that are reliable at predicting where and when initiation will occur. While the models can answer a great many questions about how the convective event will unfold, the key to specific issues often lies in the subjective analysis and diagnosis, not in the dynamical model solution.

13.6. ENSEMBLE FORECASTING. During the 1990s and 2000s, supercomputer processing speeds began to eclipse the diminishing returns from increasingly sophisticated atmospheric models. One solution that has gained acceptance is to execute a model run numerous times, each with slightly varied parameters in the initialized data. The premise here is that the correct solution is theoretically contained somewhere within the spectrum of the ensemble members. The ensemble mean indicates the most probable solution. However in reality only a few of the members have the correct solution, if at all. ¤

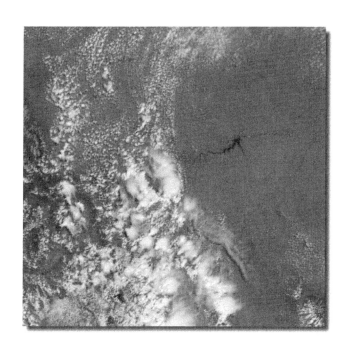

APPENDIX

WSR-88D DESCRIPTION

NEXRAD (Next Generation Weather Radar) is the primary weather radar program of the United States, developed between 1979 and 1991 as a tri-agency partnership between the National Oceanic and Atmospheric Administration, the Federal Aviation Administration, and the Department of Defense. It relies on a network of 159 radar units named WSR-88D (Weather Service Radar model 88D). It replaces the WSR-57 and WSR-74C network that was heavily used since 1959.

The WSR-88D was designed for flexibility and relies upon a modular architecture that allows for the system to be upgraded over the years. It is comprised of several subsystems.

In Canada, the weather agency began with a network of 16 weather radars and 4 Doppler radars. Due to increasing demand for severe weather, hydrology, and environmental data this was upgraded starting in 1997 with 30 Doppler radars. These are not described in detail here since Environment Canada largely restricts the public distribution of radar images, but the basic principles for interpretation presented elsewhere in this book still apply.

WSR-88D design

The WSR-88D is comprised of three primary subsystems. They are as follows:

1. RADAR DATA ACQUISITION (RDA). This is the antenna, transmitter, and computer system which is responsible for transmitting the radar pulse and feeding the RPG (below) a set of base data: reflectivity, velocity, and spectrum width. The antenna is 28 feet in diameter inside a 39-foot diameter fiberglass radome, and features a beam width of 0.93°. Some forecasters use the term "RDA" to refer generically to the radar site as a whole.

2. RADAR PRODUCT GENERATOR (RPG). This is a computer subsystem located at the radar site which ingests all of the RDA data. It not only produces graphical maps but runs all of the storm algorithms. It transmits these products to the outside world.

3. MASTER SWITCH CONTROL FUNCTION (MSCF), formerly the UNIT CONTROL POSITION (UCP). Every WSR-88D site has one MSCF workstation, always located within a weather office. It primarily runs the HCI (Human Control Interface), which is used to change scan modes, start the backup generator, monitor problems with the RDA and RPG, and many other functions. The older UCP was comprised of a cumbersome alphanumeric terminal and was retired in 2001.

4. PRINCIPAL USER DISPLAY (PUP). Until the late 1990s, all radar data used in an official capacity was displayed on a Principal User Processor (PUP) workstation comprised of dual CRT monitors, a tablet, and alphanumeric terminal. The specialized user interface and 19-inch CRTs rapidly became outdated and display capabilities were eventually integrated into the National Weather Service's AWIPS system. Two similar workstations are used inhouse by the Federal Aviation Administration. Radar products are also disseminated to the general public in binary form via NOAA data gateways.

Volume coverage pattern (VCP)

The WSR-88D has historically used four scan strategies, which are combinations of radar elevations, scan speeds, and pulse lengths that are optimum for the meteorological situation at hand. These strategies are known as Volume Coverage Patterns (VCP). During much of the 1990s, 11, 21, 31, and 32 were the only ones in use, with the former two being "precipitation modes" and the latter two being "clear air modes".

It is important for forecasters who use WSR-88D products to be familiar with the VCP in effect as this influences the quality and interpretation of radar data. Note that elevation in radar meteorology always refers specifically to the tilt of the radar beam from the horizontal plane. It does not refer to altitude MSL or AGL unless specifically qualified as such.

1. VCP 11 — STORM SCAN. This is a storm scan VCP that takes 5 minutes and uses 14 elevations between 0.5° and 19.5°. It offers the greatest resolution at higher elevations, making it the best choice for storms close to the radar site.

2. VCP 211 — VCP 11 WITH SPLIT CUT. This is the same as VCP 11 (q.v.) but adds the SZ-2 range-folding mitigation algorithm.

3. VCP 12 — LOW ELEVATION STORM SCAN. Identical to VCP 11 (q.v.) but completes in 4.5 minutes and is *biased heavily toward the lower elevations*, thus there are more low-elevation scans with better vertical resolution but fewer high-elevation scans. The higher antenna rotation rate also degrades the data quality somewhat. VCP 12 is best used for storms at distant range from the radar (i.e. over 60 nm).

4. VCP 212 — VCP 12 WITH SZ-2 ALGORITHM. This is the same as VCP 12 (q.v.) but adds the SZ-2 range-folding mitigation algorithm.

5. VCP 21 — SHALLOW STORM SCAN. This is a storm scan VCP that takes 6 minutes and scans 9 elevations between 0.5° and 19.5°. There are many gaps above 5°, which presents problems with nearby storms. VCP 21 offers few advantages over VCP 11.

6. VCP 121 — VCP 21 WITH VARIABLE PRF. Similar to VCP 21, scanning 9 elevations between 0.5° and 19.5°, but takes 5.5 instead of 6 minutes. The chief advantage of VCP 121 is that it uses a variable PRF to minimize velocity aliasing problems.

7. VCP 221 — VCP 21 WITH SPLIT CUT. This is the same as VCP 21 (q.v.) but adds the SZ-2 range-folding mitigation algorithm.

8. VCP 31 — LONG-PULSE CLEAR AIR MODE. A clear air VCP that takes 10 minutes and scans 5 elevations between 0.5° and 4.5°. VCP 31 is more prone to aliasing problems than VCP 32.

9. VCP 32 — SHORT-PULSE CLEAR AIR MODE. A clear air VCP that takes 10 minutes and scans 5 elevations between 0.5° and 4.5°. It is more prone to range folding problems than VCP 31.

Super resolution radar

Beginning in May 2008, WSR-88D sites began offering Super Resolution radar. This improves radial resolution from 1.0 to 0.5 deg and range resolution from 1 km (0.54 nm) to 0.25 km (0.13 nm) for all *base products*. It is not available on all VCP elevations, especially at higher ones. The improvements are significant and are already available to the public. Forecasters should make full efforts to use these enhanced products.

Quick Reference VCP Comparison Table for RPG Operators

February 2007

Slices	Tilts	VCP	Time*	Usage	Limitations
19.5° 16.7° 14.0° 12.0° 10.0° 8.7° 7.5° 6.2° 5.3° 4.3° 3.4° 2.4° 1.5° 0.5°	14	11	5 mins	Severe and non-severe convective events. Local 11 has Rmax=80nm. Remote 11 has Rmax=94nm.	Fewer low elevation angles make this VCP less effective for long-range detection of storm features when compared to VCPs 12 and 212.
19.5° 15.6° 12.5° 10.0° 8.0° 6.4° 5.1° 4.0° 3.1° 2.4° 1.8° 1.3° 0.9° 0.5° 0.0°	14	211	5 mins	Widespread precipitation events with embedded, severe convective activity (e.g. MCS, hurricane). Significantly reduces range-obscured V/SW data when compared to VCP 11.	All Bins clutter suppression is NOT recommended. PRFs are not editable for SZ-2 (Split Cut) tilts.
	14	12	4½ mins	Rapidly evolving, widespread severe convective events. Extra low elevation angles increase low-level vertical resolution when compared to VCP 11.	High antenna rotation rates decrease the effectiveness of clutter filtering, increase the likelihood of bias, and slightly decrease accuracy of the base data estimates.
	14	212	4½ mins	Rapidly evolving, widespread severe convective events (e.g. squall line, MCS). Increased low-level vertical resolution compared to VCP 11. Significantly reduces range-obscured V/SW data when compared to VCP 12.	All Bins clutter suppression is NOT recommended. PRFs are not editable for SZ-2 (Split Cut) tilts. High antenna rotation rates decrease the effectiveness of clutter filtering, increase the likelihood of bias, and slightly decrease accuracy of the base data estimates.
	9	21	6 mins	Non-severe convective precipitation events. Local 21 has Rmax=80nm. Remote 21 has Rmax=94nm.	Gaps in coverage above 5°.
19.5° 14.6° 9.9° 6.0° 4.3° 3.4° 2.4° 1.5° 0.5° 0.0°	9	121	6 mins	VCP of choice for hurricanes. Widespread stratiform precipitation events. Significantly reduces range-obscured V/SW data when compared to VCP 21.	PRFs are not editable for any tilt. Gaps in coverage above 5°.
	9	221	6 mins	Widespread precipitation events with embedded, possibly severe convective activity (e.g. MCS, hurricane). Further reduces range-obscured V/SW data when compared to VCP 121.	All Bins clutter suppression is NOT recommended. PRFs are not editable for SZ-2 (Split Cut) tilts. Gaps in coverage above 5°.
	5	31	10 mins	Clear-air, snow, and light stratiform precipitation. Best sensitivity. Detailed boundary layer structure often evident.	Susceptible to velocity dealiasing failures. No coverage above 5°. Rapidly developing convective echoes aloft might be missed.
4.5° 3.5° 2.5° 1.5° 0.5° 0.0°	5	32	10 mins	Clear-air, snow, and light stratiform precipitation.	No coverage above 5°. Rapidly developing convective echoes aloft might be missed.

NATIONAL WEATHER SERVICE / WARNING DECISION TRAINING BRANCH

DIAGNOSTIC VARIABLES

This section summaries the key diagnostic variables used in severe weather forecasting. Other historical indices are not presented here and may be found in *Weather Forecasting Red Book*.

1. SHOWALTER STABILITY INDEX (SI, SSI). The SI is one of the oldest diagnostic variables in existence, having been developed in the 1940s by Showalter. It is not as common in severe weather forecasting as it was in the 1950s and 1960s, but is still encountered today. It is defined as:

$SI = T_{E500} - T_{L500}$

where T_{E500} is the environmental temperature at 500 mb and T_{L500} is the final temperature of a parcel lifted from the 850 mb level to 500 mb. In short, this compares the temperature of a parcel with its environment. Note that the parcel selection is that from 850 mb, not the surface or a mixed layer, so it is strongly influenced by the depth of the moist layer.

2. LIFTED INDEX (LI). The Lifted Index was developed by SELS forecaster Joseph Galway in the early 1950s and published in 1956. It is very similar to the SI in comparing a parcel lifted to 500 mb, except instead of using a parcel starting at the 850 mb level, it uses a parcel starting at the earth's surface using the forecast temperature and dewpoint.

$LI = T_{E500} - T_{L500}$

Values of below zero are considered significant for thunderstorm potential, and values of -5 or below suggest severe weather.

3. CONVECTIVE AVAILABILITY OF POTENTIAL ENERGY (CAPE). In the 1990s, CAPE became the favored measure of instability by severe storm forecasters.

$CAPE = \int (\alpha_{lp} - \alpha) dp$

where $\int$ indicates an integral from the LFC to the EL.

CAPE by itself is very effective at distinguishing non-thunderstorm environments from thunderstorm environments. Its skill in distinguishing the possibility of supercells from non-supercells is marginal, however, with 500-1000 J·kg^{-1} being a lower end ML-CAPE value for supercells. For forecasting tornadoes it is not reliable. Combined parameters have been found to work much better for predicting these occurrences than CAPE alone.

4. BRN SHEAR. BRN Shear is defined as U^2, where U is the difference in the density-weighted mean winds, in m·s^{-1}, between the 0-6 km AGL layer and the 0-500 m AGL layer.

5. BULK RICHARDSON NUMBER (BRN). Research during the 1980s showed the BRN to have some use as a discriminator of storm type. Rather than using both buoyancy and shear together to amplify the possibility of rotating storms, BRN proposed that rotating storms will develop if there is a balance between buoyancy and shear. It is calculated as follows:

$$BRN = CAPE / 0.5(U^2)$$

where U^2 is the BRN Shear (q.v.)

Studies published in the 1980s found that BRN of 10 to 45 favors supercells. Values above 45 are associated with ordinary or multicell storms while values below 10 are associated with detrimental shear profiles. Additional studies performed in the 1990s, however, have uncovered a lot of overlap between storm types. Caution is advised in using BRN because very high CAPE situations can and do produce tornadoes, and in these situations the BRN will be out of the accepted range of values for supercells.

6. ENERGY-HELICITY INDEX (EHI). The EHI (Hart and Korotky 1991) is an index that reconciles instability with shear in order to forecast tornado potential.

$$EHI = CAPE \cdot SRH_{(0-3)} / 160,000$$

where $SRH_{(0-3)}$ is the SRH in the 0-3 km AGL layer. A revised version (Rasmussen 2003) uses the 0-1 km AGL layer and has been found to be a better predictor of tornado potential.

By definition, EHI values above 1 indicate possible tornadoes and above 2 indicate strong tornadoes. EHI however depends on proper selection of the SRH layer and a representative and properly-modified sounding. A study found that half of all ordinary thunderstorms develop in an environmental EHI of 0 to 0.34, while half of all nontornadic supercells develop with an EHI range of 0.20 to 1.46; furthermore, half of all tornadic supercells were associated with a range of 0.42 to 2.87 (Rasmussen and Blanchard 1998).

7. VORTICITY GENERATION PARAMETER (VGP). The VGP relates the storm inflow to the tilting of vorticity into the vertical. It is as follows:

$$VGP = S \times (CAPE)^{0.5}$$

where S equals the mean shear (hodograph length divided by layer depth)

VGP has shown some skill at differentiating supercell environments from non-supercell environments.

8. SIGNIFICANT TORNADO PARAMETER (STP). The STP index was created as a means of identifying environments that are significantly tornadic. The definition of STP is:

$$STP = (CAPE_{ML}/1000)(s_6/20)(SRH_1/100)([2000-LCL_{ML}]/1500)$$

where s_6 is the shear magnitude (m·s^{-1}) from the surface to 6 km AGL, SRH_1 is the SRH in the 0-1 km AGL layer, and LCL is in meters. It uses a multiplicative product of shear and instability, making it a close cousin of EHI, but it uses two shear parameters, factors in layer humidity, and scales each component. Since it is a product-based index, if CAPE or shear diminish to zero, STP also diminishes to zero. Values of above 1 are associated with strong tornadoes given discrete storm modes, while those below 1 are associated with nontornadic storms.

9. SUPERCELL COMPOSITE PARAMETER (SCP). The SCP was designed to identify environments supportive of right-moving supercells. It is defined as:

$$SCP = (CAPE_{MU}/1000)(SRH_3/100)(U/40)$$

where SRH_3 is the SRH in the 0-3 km AGL layer and U is the BRN Shear number (see the definition earlier in this section). Values of below 2 are associated with nonsevere storms, while those above 2 are associated with supercells given discrete storm modes.

10. ENHANCED STEERING POTENTIAL (ESP). Defined as:

$$ESP = (l2 - 7)(CAPE_{ML3}/50)$$

11. MCS INDEX. The MCS Index (Jirak and Cotton 2007) is a diagnostic index that attempts to determine the likelihood that convection will develop into an MCS. The definition is:

$$MCSI = (LI + 4.4)/-3.3 + (S - 11.5)/5 + (TA_{700} - 4.5 \times 10^{-5})/7.3 \times 10^{-5}$$

where LI is the lifted index, S is the bulk shear in $m \cdot s^{-1}$ between 0 and 3 km, and TA_{700} is the 700 mb temperature advection in units of $10^{-5} K \cdot s^{-1}$. It should only be used in areas where thunderstorms are already expected.

It can be seen that the MCSI is comprised of three terms. First is a stability term, which contributes positively to the MCSI when the LI is below -7.7. Second is a low-level bulk-shear term, which contributes positively to the MCSI when the shear exceeds 16.5 m/s (32 kt). Finally is a mid-level temperature advection, which contributes positively to the MCSI when at least $11.8 \times 10^{-5} K \cdot s^{-1}$ of warm air advection is taking place, i.e. about half a Celsius degree per hour.

MCSI values of below -1.5 are considered unfavorable for MCS development; between -1.5 and 0 are considered marginal; between 0 and 3 are considered favorable; and values exceeding 3 are considered very favorable.

Hodograph

This diagram may be freely reproduced without restriction.

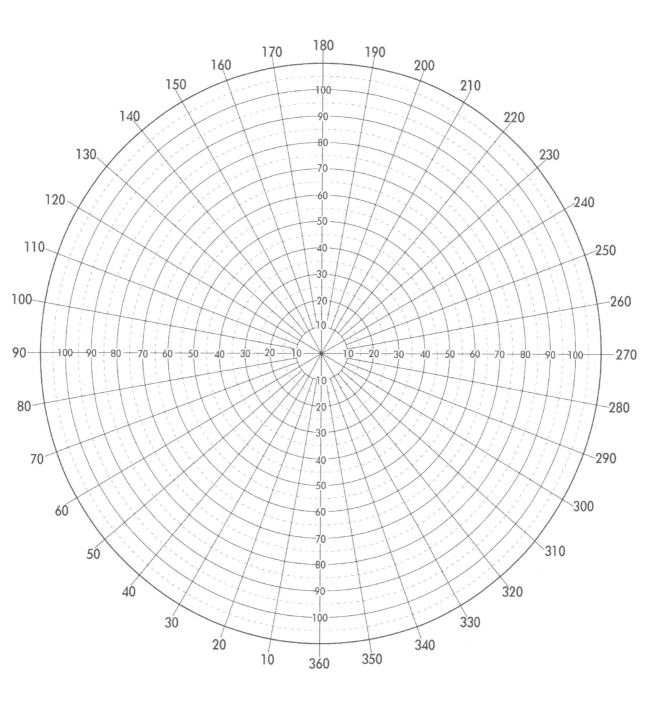

Skew-T log P Diagram

This diagram may be freely reproduced without restriction.

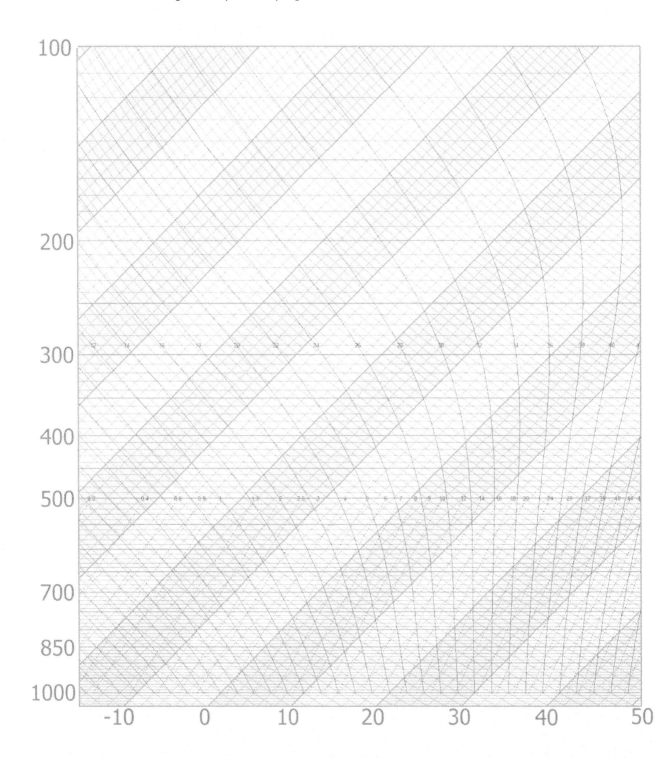

REFERENCES

The volume of literature on severe weather is vast and I have strived to focus on those that have a bearing on operational forecasting. If I have omitted a noteworthy paper, it is either unintentional or was felt to have too strong of a research emphasis for inclusion here. This text focuses on fundamentals so case studies are generally omitted.

Agee, Ernest and Erin Jones, 2009: Proposed conceptual taxonomy for proper identification and classification of tornado events. *Wea. and Forecasting*, 24, 609-617.

Andrus, C. G., 1929: Notes on line squalls. *Mon. Wea. Rev.*, 57, 94-96.

Atkins, Nolan T. and Michael St. Laurent, 2009: Bow echo mesovortices: Part I: Processes that influence their damage potential. *Mon. Wea. Rev.*, 137, 1497-1513.

Atkins, Nolan T. and Michael St. Laurent, 2009: Bow echo mesovortices: Part II: Their genesis. *Mon. Wea. Rev.*, 137, 1514-1532.

Banacos, P. C., and D. M. Schultz, 2005: The use of moisture flux convergence in forecasting convective initiation: Historical and operational perspectives. *Wea. Forecasting*, 20, 351–366.

Bates, Ferdinand C., 1967: A Major Hazard to Aviation near Severe Thunderstorms. *Aviation Safety Monograph 1*, Department of Geosciences, St. Louis University, 36 pp. Established the flanking line concept.

Beebe, Robert G. and Ferdinand C., 1955: A mechanism for assisting in the release of convective instability. *Mon. Wea. Rev.*, 83, 1-10. The first paper to explore conceptual models of jet streak quadrants for breaking the cap.

Bentley, Mace L., Thomas L. Mote, and Stephen F. Byrd, 1998: A synoptic climatology of derecho producing mesoscale convective systems: 1986-1995. *19th Conf. on Severe Local Storms*, Minneapolis, 5-8. This article summarizes the distribution of derecho events across the U.S.

Bernardet, Ligia R. and William R. Cotton, 1998: Multiscale evolution of a derecho-producing mesoscale convective system. *Mon. Wea. Rev.*, 126, 2991-3015. Proposed internal gravity waves in an MCS as augmenting damaging downburst events.

Blanchard, David O., 1998: Assessing the vertical distribution of convective available potential energy. *Wea. Forecasting*, 13, 870–877.

Bluestein, Howard and Carlton R. Parks, 1983: A synoptic and photographic climatology of low-precipitation severe thunderstorms in the Southern Plains. *Mon. Wea. Rev.*, 111, 2034-2046. Formally defined the LP supercell and identified characteristic features.

_____ and Michael H. Jain, 1985: Formation of mesoscale lines of precipitation: Severe squall lines in Oklahoma during the spring. *J. Atmos. Sci.*, 42, 1711-1732. Devised a classification scheme for squall lines.

_____ and Morris L. Weisman, 2000. The interaction of numerically simulated supercells. *Mon. Wea. Rev.*, 128, 3128-3149. This study explored the interplay of orientation of deep-layer shear vectors and boundaries on storm morphology.

Bright, David R. and Darren M. McCollum, 1998: Monitoring Gulf of California moisture surges with GOES-9 data and the WSR-88D at Yuma, Arizona. Preprints, 16th Conference on Weather Analysis and Forecasting, Amer. Met. Soc., 50-52.

Brock, F. V., and S. J. Richardson, 2001: *Meteorological Measurement Systems*. Oxford University, 290 pp.

Brooks, Edward M., 1949: The tornado cyclone. *Weatherwise*, 2, 32-33. One of the first articles to describe a "tornado cyclone" in March 19, 1948 in St Louis. It must be understood that early-era terminology describing "tornado cyclones" are misnomers and actually refer to the storm-scale mesocyclone.

Brooks, Harold E., and Charles A. Doswell III, 1993: Extreme winds in high-precipitation supercells. *Preprints, 17th Conf. on Severe Local Storms*, St. Louis, 173-177. Demonstrated the ability of HP storms as defined by Moller et al 1988, 1990 to produce extreme straight-line winds.

_____, 2004: On the relationship of tornado path length and width to intensity. *Wea. and Forecasting*, 19, 310-319. Quantifies observed damage path length and width to rated intensity.

Browning, Keith A., 1962: Cellular structure of convective storms. *Meteorological Magazine*, 91, 341-350. The first identification of a supercell storm in a complex of storms in England in July 1959.

_____ and R. J. Donaldson, 1963: Airflow and structure of a tornadic storm. *J. Atmos. Sci.*, 20, 533-545. Showed evidence of overhang, a vault, and what was later defined as the WER and BWER.

_____, 1964: Airflow and precipitation trajectories within severe local storms which travel to the right of the winds. *J. Atmos. Sci.*, 21, 634-639. This paper established the term "supercell".

_____, 1965: A family outbreak of severe local storms - A comprehensive study of the storms in Oklahoma on 26 May 1963. *Part I. Air Force Cambridge Research Lab, Special Report No. 32*, 346 pp. Perhaps the earliest paper to identify a supercell "collapse" phase.

Brooks, H. E, C. A. Doswell III, and J. Cooper, 1994: On the environments of tornadic and non-tornadic mesocyclones. *Wea. Forecasting*, 10, 606-618.

————, J. W. Lee, and J. P. Craven, 2003: The spatial distribution of severe thunderstorm and tornado environments from global reanalysis data. *Atmos. Res.*, 67–68, 73–94.

Brunk, I. W., 1953: Squall lines. *Bull. Amer. Meteor. Soc.*, 34, 1-9. Argued against the prevailing view that squall lines were responsible for most tornadoes.

Brunner, Jason C., Steven A. Ackerman, and A. Scott Bachmeier, 2007: A quantitative analysis of the enhanced-V feature in relation to severe weather. *Wea. Forecasting*, 22, 853-872. Analysis and verification of enhanced-V cases in 2003 and 2004.

Bunkers, M. J., B. A. Klimowski, J. W. Zeitler, Richard L. Thompson, and M. L. Weisman, 2000: Predicting supercell motion using a new hodograph technique. *Wea. Forecasting*, 15, 61-79. Replaces older storm motion methods with a new hodograph technique, which is now in widespread use.

Bunkers, Matthew J. and Leslie R. Lemon, 2007: Can spectrum width help identify large hail? *Preprints, NWA 32nd Annual Meeting*, Reno, NV, 13-18 October 2007.

Carr, Frederick H. and James P. Millard, 1985: A composite study of comma clouds and their association with severe weather over the Great Plains. *Mon. Wea. Rev.*, 113, 370-387. Discusses aspects of dry-slot and cold-core convection.

Chisholm, Alexander J., 1970: *Alberta Hailstorms: A Radar Study and Model*. Ph.D. thesis, Dept. of Meteorology, McGill University, Montreal, 237 pp. Establishes the weak echo region.

Cohen, Ariel E., Michael C. Coniglio, Stephen F. Corfidi, and Sarah J. Corfidi, 2007: Discrimination of mesoscale convective system environments using sounding observations. *Wea. Forecasting*, 22, 1045-1062. Examined sounding properties in different types of MCS environments.

Colman, B. R., 1990: Thunderstorms above frontal surfaces in environments without positive CAPE; Part 1 and 2. *Mon. Wea. Rev.*, 118, 1103-1122 and 1123-1144. One of the first detailed examinations of elevated storms.

Colquhoun, J. R., 1987: A decision tree method of forecasting thunderstorms, severe thunderstorms and tornadoes. *Wea. Forecasting*, 2, 337-345. Introduced a flowchart method of identifying severe thunderstorm environments.

Craven, Jeffrey P. and Ryan E. Jewell, 2002: Comparison between Observed Convective Cloud-Base Heights and Lifting Condensation Level for Two Different Lifted Parcels. *Wea. Forecasting*, 17, 885-

890. Identifies and speculates on the value of mixed layer parcels in convective diagnostics.

Darkow, G. L. and J. C. Roos, 1970: Multiple tornado producing thunderstorms and their apparent cyclic variations in intensity. *Preprints, 14th Conf. on Radar Meteorology*, Tucson, Amer. Met. Soc., 305-308. Established the concept of cyclic supercells.

Davies-Jones, Robert P., 1984: Streamwise vorticity: The origin of updraft rotation in supercell storms. *J. Atmos. Sci.*, 41, 2991-3006. Perhaps the first paper to discuss streamwise vorticity as a prerequisite for storm rotation.

————, 2002: Linear and non-linear propagation of supercell storms. *J. Atmos. Sci.*, 59, 3178-3205. Discussed aspects of supercell propagation and shear depths. Also see counterpoints in Rotunno et al 2003.

————, R., D. Burgess, and M. Foster, 1990: Test of helicity as a tornado forecast parameter. *Preprints, 16th Conf. on Severe Local Storms*, Kananaskis Park, Alta., Canada, Amer. Meteor. Soc., 588–592. Formally introduced storm-relative helicity as an operationally-useful diagnosis of streamwise vorticity.

Davies, Jonathan M., 2006: Tornadoes in environments with small helicity and/or high LCL heights. *Wea. Forecasting*, 21, 579-594.

————, 2006: Tornadoes with cold core 500-mb lows. *Wea. Forecasting*, 21, 1051-1062.

Doswell, C. A. III, 1977: Obtaining meteorologically significant surface divergence fields through the filtering property of objective analysis. *Mon. Wea. Rev.*, 105, 885–892.

————, 1986: Short range forecasting. *Mesoscale Meteorology and Forecasting*. Peter S. Ray, Ed., Amer. Meteor. Soc., 689–719. An excellent overview of nowcasting.

————, 1988: Comments on "An improved technique for computing the horizontal pressure-gradient force at the earth's surface." *Mon. Wea. Rev.*, 116, 1251–1254.

————, 2004: Weather forecasting by humans—Heuristics and decision making. *Wea. Forecasting*, 19, 1115–1126. A rare look into the mental process of creating forecast based on sound scientific principles.

————, and E. N. Rasmussen, 1994: The effect of neglecting the virtual temperature correction on CAPE calculations. *Wea. Forecasting*, 9, 625–629.

————, and P. M. Markowski, 2004: Is buoyancy a relative quantity? *Mon. Wea. Rev.*, 132, 853–863.

————, H. E. Brooks, and R. A. Maddox, 1996: Flash flood forecasting: An ingredients-based methodology. *Wea. Forecasting*, 11, 560–581.

————, D. V. Baker and C. A. Liles 2002: Recognition of negative mesoscale factors for severe weather potential: A case study. *Wea. Forecasting*, 17, 937–954.

Edwards, Roger and Richard L. Thompson, 1998: Nationwide comparisons of hail size with WSR-88D vertically integrated liquid water and derived thermodynamic sounding data. *Wea. Forecasting*, 13, 277-285. Casts doubt on VIL as an effective hail indicator.

————, 2006: Supercells of the Serranias del Burro (Mexico). *23rd Conference on Severe Local Storms*, St. Louis. Examines unusually strong storms in the region west of and near Del Rio, Texas.

Evans, Jeffry S. and Charles A. Doswell III, 2001: Examination of derecho environments using proximity soundings. *Wea. and Forecasting*, 16, 329-342. Reviews instability and shear profiles with a set of 1983-93 derecho storms.

Fankhauser, J. C., 1976: Structure of an evolving hailstorm: Part II: Thermodynamic structure and airflow in the near environment. *Mon. Wea. Rev.*, 104, 576-587.

Fawbush, Ernest F. and Robert C. Miller, 1953: A method for forecasting hailstone size at the earth's surface, *Bull. Amer. Meteor. Soc.*, 34, 235-244. Sets forth the earliest hail forecasting algorithm. Omitted from AMS holdings online.

Foster, Donald S. and Ferdinand C. Bates, 1956. A hail size forecasting technique. *Bull. Amer. Meteor. Soc.*, 37, 135-141. Proposed an early hail forecast method. Not in AMS holdings online, but is reviewed in P. W. Leftwich, Preprints 10th Conf. *Wea. Forecasting* and Analysis, 1984, 525-528.

Fujita, Theodore, 1958: Mesoanalysis of the Illinois tornadoes of 9 April 1953. *J. Meteor.*, 15, 288-296. Established the hook echo as being associated with rotation.

_____, 1960: A Detailed Analysis of the Fargo Tornadoes of June 20, 1957. *Research paper to the U.S. Weather Bureau, No. 42*, University of Chicago, 67 pp. Identified the wall cloud.

_____, 1965: Formation and steering mechanisms of tornado cyclones and associated hook echoes. *Mon. Wea. Rev.*, 93, 67-78. Proposes the Magnus effect as an ingredient in deviant movement.

_____ and H. R. Byers, 1977: Spearhead echo and downburst in the crash of an airliner. *Mon. Wea. Rev.*, 105, 129-146. Introduced the downburst to the severe storm forecast community.

_____, 1978: Manual of downburst identification for project NIMROD. *Satellite and Mesometeorology Res. Paper 156*, Dept. of Geophysical Sciences, Univ. of Chicago, 104 pp. One of the first papers to identify the bow echo.

_____, 1981: Tornadoes and downbursts in the context of generalized planetary scales. *J. Atmos. Sci.*, 38, 1511-1534. Alongside the 1975 Orlanski paper this one proposed a modern system of scales of motion.

_____, 1986: Mesoscale classifications: their history and their application to forecasting. *Mesoscale Meteorology and Forecasting*, Peter S. Ray, Ed., Amer. Meteor. Soc., 18-35.

Gallus, William A. Jr., Nathan A. Snook, and Elise V. Johnson, 2008: Spring and summer severe weather reports over the Midwest as a function of convective mode: A preliminary study. *Wea. and Forecasting*, 23, 101-113. Outlined a scheme for classifying radar echoes.

Galway, Joseph G., 1956: The lifted index as a predictor of latent instability. *Bull. Amer. Meteor. Soc.*, 43, 528–529. SELS forecaster Joseph Galway establishes and defines the lifted index, one of the first diagnostic variables in storm forecast use.

_____, 1977: Some climatological aspects of tornado outbreaks. *Mon. Wea. Rev.*, 105, 477-484. Identified grouping patterns of tornado paths.

Garcia, C., Jr., 1994: *Forecasting snowfall using mixing ratios on an isentropic surface—An empirical study*. NOAA Tech. Memo. NWS CR-105, PB 94-188760 NOAA/NWS, 31 pp. [Available from NOAA/National Weather Service Central Region Headquarters, Kansas City, MO 64106-2897.]

George, J. J., 1960: *Weather Forecasting for Aeronautics*. Academic Press, 673 pp.

Glickman, T. S., Ed., 2000: *Glossary of Meteorology*. 2nd ed., Amer. Meteor. Soc., 855 pp.

Goetsch, E. H., 1988: Forecasting cold core severe weather outbreaks. *Preprints, 15th Conference on Severe Local Storms*, Baltimore, Amer. Met. Soc., 362-365. One of the first papers to deal with severe weather from cold core convection.

Haltiner, G. J., and R. T. Williams, 1980: *Numerical Prediction and Dynamic Meteorology*, 2nd ed. John Wiley & Sons, 477 pp.

Hart, J. A., and W. Korotky, 1991: *The SHARP workstation v1.50 users guide*. National Weather Service, NOAA, U.S. Department of Commerce, 30 pp. [Available from NWS Eastern Region Headquarters, 630 Johnson Ave., Bohemia, NY 11716.] Formally defines the Energy-Helicity Index (EHI).

Herlofson, Nicolai, 1947: The T, log p Diagram with Skew Coordinate Axes. *Meteorologiske Annaler*, Band II, No. 10, 311-342, Met. Inst., Oslo. This paper by Norwegian meteorologist Nicolai Herlofson introduced the skew-T log-p diagram.

Hitschfeld, W., 1960: The Motion and Erosion of Convective Storms in Severe Vertical Wind Shear. *J. Meteor.*, 17, 270-282. Originally published in 1959 as a McGill University thesis, this paper established the role of shear in severe weather and proposed separation of the updraft and downdraft.

Houston, Adam L., Richard L. Thompson, and Roger Edwards, 2008: The optimal bulk wind differential depth and the utility of the upper-tropospheric storm-relative flow for forecasting. *Wea. and Forecasting*, 23, 825-837.

Humphreys, William J., 1914: The thunderstorm and its phenomena. *Mon. Wea. Rev.*, 42, 348-380. An excellent overview of what was known about thunderstorm forecasting in the 1910s.

Hutchinson, Todd A. and Howard B. Bluestein, 1998: Prefrontal wind-shift lines in the Plains of the United States. *Mon. Wea. Rev.*, 126, 141-166.

Jirak, Israel L. and William R. Cotton, 2007: Observational analysis of the predictability of mesoscale convective systems. *Wea. Forecasting*, 22, 813-838. Establishes an MCS forecasting index. In *Wea. Forecasting*, Vol. 24, 351-355, Bunkers warned that results could be biased strongly in gridded datasets by the temperature advection term.

Johns, Robert H. and W. D. Hirt, 1983: The derecho... a severe weather producing convective system. *Preprints, 13th Conf. Severe Local Storms*, Tulsa, Amer. Met. Soc., 178-181. One of the first papers to establish the modern-day recognition of the derecho.

_____, Kenneth W. Howard, and Robert A Maddox, 1990: Conditions associated with long-lived DERECHOS - An examination of the large-scale environment. *Preprints, 16th Conf. on Severe Local Storms*, Kananaskis Park, 408-412. A further review of conditions favoring derecho development.

_____, and Charles A. Doswell III, 1992: Severe local storms forecasting. *Wea. and Forecasting*, 7, 588–612. An excellent review of operational forecasting techniques in the early 1990s.

Johnson, Brenda C., 1983: The heat burst of 29 May 1976. *Mon. Wea. Rev.*, 111, 1776-1792. Formally established the first accepted theory for heat burst events.

Johnston, Edward C., 1982: Mesoscale vorticity centers induced by mesoscale convective complexes. *9th Conf. on Wea. Forecasting and Analysis*, Seattle, 196-200. Perhaps the first published paper to identify the mesoscale convective vortex (MCV).

Klemp, J. B., R. Rotunno, and M. L. Weisman, 1985: Numerical simulations of squall lines in two and three dimensions. *Preprints, 14th Conf. Severe Local Storms*, Indianapolis, 179-182.

Koch, Steven E. and John McCarthy, 1982: The Evolution of an Oklahoma Dryline, Part II: Boundary-Layer Forcing of Mesoconvective Systems. *J. Atmos. Sci.*, 39, 237-257.

Kuchera, Evan L. and Matthew D. Parker, 2006: Severe convective wind environments. *Wea. Forecasting*, 21, 595-612. A review of diagnostic parameters for forecasting strong outflow.

Lemon, Leslie R., 1980: *Severe thunderstorm radar identification techniques and warning criteria*. NOAA Tech. Memo. NWS NSSFC-3, 60 pp. Established some early guidelines for identifying large hail and other features.

_____, 1998: Updraft identification with radar. *19th Conf. on Severe Local Storms*, Minneapolis, 709-712.

_____, 2009: Supercell collapse. *34th Conf. on Radar Meteorology*, Williamsburg VA.

Lewis, William and Porter J. Perkins, 1953: Recorded pressure distribution in the outer portion of a tornado vortex. *Mon. Wea. Rev.*, 81, 379-385. Identified a storm-scale mesocyclone associated with a Ohio tornado.

Livingston, Richard L., 1983: *On the Subsynoptic Pre-Tornado Surface Environment*. Ph.D. dissertation, University of Missouri, Columbia, 70 pp.

Maddox, Robert A., 1976: An evaluation of tornado proximity wind and stability data. *Mon. Wea. Rev.*, 104, 133-142. Analyzed 250 soundings near tornado events, proposing a 30 degrees right / 75% of mean flow rule for deviant motion based on research from Marwitz, Fankhauser, and Haglund.

Maddox, Robert A., 1980: Mesoscale convective complexes. *Bull. Amer. Meteor. Soc.*, 61, 1374-1387. Established the definition of the mesoscale convective complex (MCC) and described its importance in forecasting.

_____, L. Ray Hoxit, and Charles F. Chappell, 1980. A Study of Tornadic Thunderstorm Interactions with Thermal Boundaries, *Mon. Wea. Rev.*, 108, 322-336. Showed how a cool boundary can provide a field of enhanced backing of winds.

_____, 2007. *Comments on the NWS Radiosonde Replacement System (RRS)*. <http://www.squidinkbooks.com/madweather/Comments.on.NWS.RRS.pdf>. A critical article on quality control issues in the current NWS upper air network.

_____ and Barry Schwartz, 2008. *Strongly Superadiabatic Lapse Rates Aloft at Six Upper-Air Sounding Sites*. <http://www.squidinkbooks.com/madweather/pdfs/Strongly%20Superadiabatic%20Lapse%20Rates%20Aloft.pdf>

Magor, Bernard W., 1958: A meso-low associated with a severe storm. *Mon. Wea. Rev.*, 86, 81-90. Established the nomenclature of the mesolow and expanded upon the current body of knowledge.

Markowski, Paul M. and Jerry M. Straka, 2000: Some observations of rotating updrafts in a low-buoyancy, highly sheared environment. *Mon. Wea. Rev.*, 128, 449-461. Reviewed a 1998 outbreak of miniature supercells in Oklahoma.

_____, 2002: Hook echoes and rear flank downdrafts: A review. *Mon. Wea. Rev.*, 130, 852-876. Demonstrated RFD descent to be a causative factor in some hook echo occurrences.

_____, Jerry M. Straka, and Erik N. Rasmussen, 2002: Direct Surface Thermodynamic Observations within the Rear-Flank Downdrafts of Nontornadic and Tornadic Supercells. *Mon. Wea. Rev.*, 130, 1692-1721. Associated rear-flank downdraft temperature characteristics with tornadic development modes.

_____ and Yvette Richardson, 2006: On the classification of vertical wind shear as directional shear versus speed shear. *Wea. Forecasting*, 21, 242-247. Discusses important aspects of shear terminology.

Marwitz, John D., August H. Auer Jr., and Donald L. Veal, 1972: Locating the Organized Updraft on Severe Thunderstorms. *J. Appl. Meteorol.*, 11, 236-238. Established the rain-free base and its link with the updraft.

Matson, R. J., and A. W. Huggins, 1980: The direct measurement of the sizes, shapes, and kinematics of falling hailstones. *J. Atmos. Sci.*, 37: 1107-25. Examines the measured characteristics of hailstones during descent.

McCann, D. W., 1983: The enhanced-V: A satellite observable severe storm signature. *Mon. Wea. Rev.*, 111, 887-894. The first major paper to describe the enhanced V.

McCarthy, John and Steven E. Koch, 1982: The evolution of an Oklahoma dryline: Part I: A meso- and subsynoptic-scale analysis. *J. Atmos. Sci.*, 39, 225-236. An early examination of the storm-scale environment with moisture convergence.

McCaul, E. W., 1991: Buoyancy and shear characteristics of hurricane tornado environments. *Mon. Wea. Rev.*, 119, 1954-1978. Was perhaps the first paper to identify the existence of miniature supercells.

McGinley, John, 1986: Nowcasting Mesoscale Phenomena. *Mesoscale Meteorology and Forecasting*. Peter S. Ray, Ed., Amer. Meteor. Soc., 657-688.

Miller, R. C., 1972: *Notes on analysis and severe storm forecasting procedures of the Air Force Global Weather Central.* Tech. Report 200(R), Headquarters, Air Weather Service, Scott Air Force Base, IL 62225, 190 pp.

Miller, D. J., and Robert H. Johns, 2000: A detailed look at extreme wind damage in derecho events, *Preprints, 20th Conf. on Severe Local Storms*, Orlando, FL, Amer. Meteor. Soc., 52-55. Ascribed some damaging MCS winds to HP supercells embedded within a line.

Moller, Alan R., Charles A. Doswell III, John McGinley, Steven Tegtmeier, and Randy Zipser, 1974: Field observations of the Union City tornado in Oklahoma. *Weatherwise*, 27, 67-77. Identified the clear slot and illustrated other key features of tornadic supercells.

_____, 1980: Mesoscale surface analysis of the 10 April 1979 tornadoes in Texas and Oklahoma. Preprints, 8th Conf. on *Wea. Forecasting*, Denver, 36-43. Documented mesoscale analysis on 4/10/79 which revealed mesolows and small-scale pressure fall waves.

_____, and Charles A. Doswell III, 1988: A proposed advanced storm spotter's training program. Preprints, 15th Conf. Severe Local Storms, Baltimore MD, Amer. Meteor. Soc., 173-177. Formally introduced the HP supercell and advocated the supercell spectrum concept rather than discrete supercell types.

_____, Charles A. Doswell III, and R. W. Przybylinski, 1990: High precipitation supercells: A conceptual model and documentation. *Preprints, 16th Conf. on Severe Local Storms*, Kananaskis Park, 52-57. Introduced the HP supercell in great detail, building upon the work of Moller et al 1988.

Monteverdi, J. P., C. A. Doswell III, and G. S. Lipari, 2003: Shear parameter thresholds for forecasting tornadic thunderstorms in northern and central California. *Wea. Forecasting*, 18, 357–370.

Moore, James T. and John P. Pino, 1990: An interactive method for estimating maximum hailstone size from forecast soundings. *Wea. and Forecasting*, 5, 508-525. Introduced a new algorithm for hail size forecasting.

_____, Glenn E. VanKnowe, and Richard C. Molinaro: A comparison of vertical motion approximations under varying jet streak curvatures. *Preprints, 16th Conf. on Severe Local Storms*, Kananaskis Park, 137-140. An excellent numerical simulation of the four-quadrant jet concept.

Murphy, A. H., 1993: What is a good forecast? An essay on the nature of goodness in weather forecasting. *Wea. Forecasting*, 8, 281–293.

Namias, Jerome and Philip F. Clapp, 1949: Confluence theory of the high tropospheric jet stream. *J. Met.*, 6, 330-336. An early paper that established the groundwork for the four-quadrant principle.

Nelson, S. P., 1976: Rear flank downdraft: A hailstorm intensification mechanism. *Preprints, 10th Conf. on Severe Local Storms*, Omaha, Amer. Met. Soc., 521-525.

Newton, C. W., 1950: Structure and mechanism of the prefrontal squall line. *J. Atmos. Sci.*, 7, 210-222. One of the earliest studies of MCS storms.

Nolen, R. H., 1959: A radar pattern associated with tornadoes. *Bull. Amer. Meteor. Soc.*, 40, 277-279. Identified the line echo wave pattern radar signature.

Orlanski, Isidoro, 1975: A rational subdivision of scales for atmospheric processes. *Bull. Amer. Meteor. Soc.*, 56, 527–530. Proposes what remains the most common system of atmospheric scales.

Panofsky, H., 1964: *Introduction to Dynamic Meteorology*. Pennsylvania State University, 243 pp.

Parker, Matthew D. and Richard H. Johnson, 2000: Organizational modes of midlatitude mesoscale convective systems, *Mon. Wea. Rev.*, 128, 3413-3436. This paper proposed a scheme for three primary modes of MCS development: TS, LS, and PS.

Peterson, Thomas C. and Imke Durre, 2002: A Climate Continuity Strategy for the Radiosonde Replacement System Transition. *8th Symposium on Integrated Observing and Assimilation Systems for Atmosphere, Oceans, and Land Surface (IAOS-AOLS)*, Seattle, 46 p. <http://www.ncdc.noaa.gov/oa/wmo/ccl/radiosonde.continuity.pdf> Discusses design issues of the RRS radiosonde.

Pettet, Crystalyne R. and Richard H. Johnson, 2003: Airflow and precipitation structure of two leading stratiform mesoscale convective systems determined from operational datasets. *Wea. Forecasting*, 18, 685-699. Highlights structures of leading stratiform MCS systems.

Pliske, Rebecca M., Beth Crandall, and Gary Klein, 2004: Competence in weather forecasting. *Psychological investigations of competence in decision making*, Cambridge University Press, ISBN 0521583063, pp. 40-68. An examination of the cognitive aspects of forecasting.

Ralph, F. M., P. J. Neiman, and T. L. Keller, 1999: Deep-tropospheric gravity waves created by leeside cold fronts. *J. Atmos. Sci.*, 56, 2986-3009. Discusses gravity waves along and ahead of fronts and their implications for initiating MCS systems.

Ramsay, H., and C. A. Doswell III, 2005: A sensitivity study of hodograph-based methods for estimating supercell motion. *Wea. Forecasting*, 20, 954–970.

Rasmussen, Erik N., and D. O. Blanchard, 1998: A baseline climatology of sounding-derived supercell and tornado forecast parameters. *Wea. Forecasting*, 13, 1148–1164.

_____, 2003: Refined supercell and tornado forecast parameters. *Wea. Forecasting*, 18, 530–535. Defines a 0-1 km version of the EHI.

_____, Jerry M. Straka, Matthew S. Gilmore, and Robert Davies-Jones, 2006: A preliminary survey of rear-flank descending reflectivity cores in supercell storms. *Wea. Forecasting*, 2006, 923-938. Also see Bluestein

_____, Erik N., M. S. Gilmore, and R. Davies-Jones, 2006: A preliminary survey of rear-flank descending reflectivity cores in supercell storms. *Wea. Forecasting*, 21, 923-928.

Rasmussen, Roy M. and Andrew J. Heymsfield, 1987: Melting and shedding of graupel and hail: Parts 1-3. *J. Atmos. Sci.*, 44, 2754-2803. This study explored aspects of hail growth using observation and modelling.

Richardson, Yvette P., 1999. *The influence of horizontal variations in vertical shear and low-level moisture on numerically simulated convective storms.* Ph.D. dissertation, School of Meteorology, University of Oklahoma, 236 pp.. Explored the preferential propagation of multicellular convection into deeper moisture as shown by numerical simulation.

Rose, Stanley F., Peter V. Hobbs, John D. Locatelli, and Mark T. Stoelinga, 2004: A 10-yr climatology relating the locations of reported tornadoes to the quadrants of upper-level jet streaks. *Wea. and Forecasting*, 19, 301-309. Establishes a positive correlation between jet streak quadrants and tornado activity.

Rotunno, Richard and J. B. Klemp, 1982: The influence of the shear-induced pressure gradient on thunderstorm motion. *Mon. Wea. Rev.*, 110, 136–151. Proposed a groundbreaking explanation for supercell propagation and why right-movers are favored.

_____ and M. L. Weisman, 2003: Comments on "Linear and nonlinear propagation of supercell storms." *J. Atmos. Sci.*, 60, 2413-2419. Discusses shear depth and supercell propagation as an opposing view to Davies-Jones 2002.

Ryzhkov, Alexander V., Terry J. Schuur, and Donald W. Burgess, 2005: Polarimetric tornado detection. *J. Appl. Met.*, 44, 557-570. Provides a review of supercell and tornado polarimetric signatures.

Sanders, Fred and C. A. Doswell III, 1995: A case for detailed surface analysis. *Bull. Amer. Meteor. Soc.*, 76, 505–521.

Schaefer, Joseph T., 1986: Severe thunderstorm forecasting: A historical perspective. *Wea. Forecasting*, 1, 164–189. An outstanding overview of storm forecasting from the 1880s through the 1980s.

_____, 1986: The dryline. *Mesoscale Meteorology and Forecasting.* Peter S. Ray, Ed., Amer. Meteor. Soc., 549-572. A benchmark article on the structure and behavior of the dryline.

Schultz, David M. and Philip N. Schumacher, 1999: The Use and Misuse of Conditional Symmetric Instability. *Mon. Wea. Rev.*, 127, 2709-2732. A fantastic overview of CSI with an operational perspective.

_____, Philip N. Schumacher, and Charles A. Doswell III, 2000: The Intricacies of Instabilities. *Mon. Wea. Rev.*, 128, 4143-4148. Reviews and clarifies terminology and definitions concerning instability.

_____, J. V. Cortinas Jr., and Charles A. Doswell III, 2002: Comments on "An operational ingredients-based methodology for forecasting midlatitude winter season precipitation." *Wea. Forecasting*, 17, 160–167.

Schmocker, Gary K., R.W. Przybylinski, and Y.J. Lin, 1996: Forecasting the initial onset of damaging downburst winds associated with a mesoscale convective system (MCS) using the mid-altitude radial convergence (MARC) signature. *Preprints, 15th Conf. on Wea. Analysis and Forecasting*, Norfolk, 306-311. Identified the MARC signature that precedes a bow echo event.

Showalter, Albert K., 1947: A stability index for forecasting thunderstorms. *Bull. Amer. Meteor. Soc.*, 34, 250–252.

Smith, Daniel L., Fred L Zuckerberg, Joseph T. Schaefer, and Glenn E. Rasch, 1986: Forecast problems: the meteorological and operational factors. *Meso-*

scale Meteorology and Forecasting. Peter S. Ray, Ed., Amer. Meteor. Soc., 36-49.

Snellman, Leonard W., 1977: Operational forecasting using automated guidance. *Bull. Amer. Meteor. Soc.,* 58, 1036-1044. Explains meteorological cancer.

_____, 1982: Impact of AFOS on operational forecasting. *9th Conf. on Wea. Analysis & Forecasting,* Seattle, 13-16. Introduces the so-called forecast funnel. Interestingly this idea is also proposed in a paper by German forecasters Fenner and Geb in the page before (p. 12) who prescribe ordering of other diagnostic parameters. The forecast funnel is sometimes erroneously ascribed to Snellman 1977.

Stout, G. E. and F. A. Huff, 1953: Radar records Illinois tornadogenesis. *Bull. Amer. Meteor. Soc.,* 34, 281-284. First evidence of the hook echo.

Tegtmeier, Steven A., 1974: *The role of the surface, sub-synoptic, low pressure system in severe weather forecasting.* M.S. thesis, University of Oklahoma, 66 pp. Establishes the importance of the dryline bulge.

Tepper, Morris, 1950: A proposed mechanism of squall lines: the pressure jump line. *J. Met.,* 7, 21-29. Weather Bureau forecaster Morris Tepper examined an indicator of tornado formation from early forecasting efforts. At the time, squall lines were believed to be associated with tornadogenesis.

Thompson, R. L., R. Edwards, J. A. Hart, K. L. Elmore, and P. Markowski, 2003: Close proximity soundings within supercell environments obtained from the Rapid Update Cycle. *Wea. Forecasting,* 18, 1243–1261.

Thompson, Richard L., Roger Edwards, and John A. Hart, 2002: Evaluation and interpretation of the supercell composite and significant tornado parameters at the Storm Prediction Center. *Preprints, 21st Conf. on Severe Local Storms,* San Antonio, TX, Amer. Meteor. Soc., J11–J14. Defines the supercell composite parameter (SCP) and significant tornado parameter (STP).

Thompson, Richard L., Cory M. Mead, and Roger Edwards, 2005: Effective storm-relative helicity and bulk shear in supercell thunderstorm environments. *Wea. and Forecasting,* 22, 102-115. Proposes the calculation of storm characteristics using a combination of the internal dynamics method and the effective layer.

Tennekes, H., and J. L. Lumley, 1972: *A first course in turbulence.* MIT Press, 300 pp.

Trapp, Robert J. and Robert Davies-Jones, 1997: Tornadogenesis with and without a dynamic pipe effect. *J. Atmos. Sci.,* 54, 113-133. An excellent explanation of tornadogenesis from 1997 which considers the dynamic pipe effect.

Trapp, Robert J. and Morris L. Weisman, 2000: Preliminary investigation of tornadogenesis within quasi-linear convective systems. *Preprints, 20th Conf. on Severe Local Storms,* Orlando, 273-276. Described mesoscale vortices along the leading edge of MCS systems.

_____, Sarah A. Tessendorf, Elaine Savageau Godfrey, and Harold E. Brooks, 2005a: Tornadoes from squall lines and bow echoes: Part I: Climatological distribution. *Wea. and Forecasting,* 20, 23-34. Examined tornado occurrence in bow echoes.

_____, Gregory J. Stumpf, and Kevin L. Manross, 2005b: A reassessment of the percentage of tornadic mesocyclones. *Wea. and Forecasting,* 20, 680-687. Uses the new WSR-88D mesocyclone detection algorithm to show that only 26% of all mesocyclones are associated with tornadoes.

Tudurí, E., and C. Ramis, 1997: The environments of significant convective events in the western Mediterranean. *Wea. Forecasting,* 12, 294–306.

Weisman, Morris L., and J. B. Klemp, 1982: The dependence of numerically simulated convective storms on vertical wind shear and buoyancy. *Mon. Wea. Rev.,* 110, 504–520. Defines the Bulk Richardson Number (BRN) and differentiates shear parameters in multicell storms from those of supercells. Showed that high bulk shear high instability environments favor bow echo modes.

_____ and J. B. Klemp, 1984: The structure and classification of numerically simulated convective storms in directionally varying wind shears. *Mon. Wea. Rev.,* 112, 2479-2498. Shows shear can occasionally enhance updrafts.

———, 1990: The numerical simulation of bow echoes. *Preprints, 16th Conf. on Severe Local Storms,* Kananaskis Park, 428-433. A detailed look at a numerically-modelled bow echo.

———, 1993: The genesis of severe, long-lived bow echoes. *J. Atmos. Sci.,* 50, 645–670. Discusses line-end vortexes on bow echo storms.

——— and Richard Rotunno, 2000: The use of vertical wind shear versus helicity in interpreting supercell dynamics. *J. Atmos. Sci.,* 57, 1452-1472. An excellent discussion of shear and helicity as it relates to supercell structures.

Wetzel, S. W., and J. E. Martin, 2001: An operational ingredients-based methodology for forecasting midlatitude winter season precipitation. *Wea. Forecasting,* 16, 156–167.

———, and ———, 2002: Reply. *Wea. Forecasting,* 17, 168–171.

World Meteorological Organization, 1956: *International Cloud Atlas, Abridged Atlas.* The International Cloud Atlas, published about every 20 years, remains the cardinal publication for educating meteorologists on cloud forms.

Zhang, J. W. and B. W. Atkinson, 1995: Stability and wind shear effects on meso-scale cellular convection. *Boundary Layer Meteorology,* 75, 263-285. This study examined the type of mesoscale convection pattern that was produced by different shear and stability profiles in an numerical model.

INDEX

Symbols

30R75 technique 149

A

accretion 117
adiabat 129
  dry 129
  mean 133
  moist 129. *See* temperature: potential
albedo 191
algorithms 9
aliasing 164
altimeter setting (ALSTG) 211
analogs 220
analysis 5, 10, 13
anvil crawler 112
anvil shadow 199
arcus 55
atomized rain 82

B

backbuilding
  MCS 75
backing 18
base reflectivity 161
base velocity 161
beam blockage 168
bear's cage 49
Bergeron-Findeisen process 117
bookend vortex 68. *See* line-end vortex
boundary
  orientation 49
bounded weak echo region (BWER) 176
bow echo 70, 179
Braham, Roscoe 4
breakdown bubble 87
break out, of precipitation 235
BRN. *See* Bulk Richardson Number (BRN, BRi)
BRN Shear 148
Browning, Keith 43
bubble high 212
bulge, dryline 219
Bulk Richardson Number (BRN, BRi) 148, 246
bulk shear. *See* shear: bulk
Bunkers method 149
BWER. *See* bounded weak echo region (BWER)
Byers, Horace 4

C

CA. *See* lightning
calvus 29
CANWARN 96
cap 137
CAPE. *See* Convective Availability of Potential Energy (CAPE)
CAPE density 143
CAPE robber 132, 140
capillatus 29
CC. *See* lightning
CCL. *See* Convective Condensation Level (CCL)
CG. *See* lightning
CINH. *See* Convective Inhibition (CINH)
CL 43
clear ice. *See* ice
climatology 221
closed cell 194
cloud streets 196
cold air funnel 91
cold core low 224
cold front 213, 223
cold pool 62, 70
collapse 28, 177
collector drops 116
collision and coalescence 116
conceptual model 11
conditional symmetric instability (CSI) 155
continuity 220
convection
  cold core 148, 224
  elevated 139
  high-based 140
Convective Availability of Potential Energy (CAPE) 107, 142, 245
Convective Condensation Level (CCL) 135
Convective Inhibition (CINH) 143
convective temperature (CT) 135
correlation coefficient 173
crosswise vorticity 21
CSI. *See* conditional symmetric instability (CSI)
CT. *See* convective temperature (CT)
cumulative shear. *See* shear: cumulative
cumulonimbus 29
cumulus 27, 29
cyclic tornadogenesis 88

D

DCAPE. *See* Downdraft Convective Availability of Potential Energy (DCAPE)
DCVZ 228
decision tree 9
Denver convergence vorticity zone (DCVZ) 228
derecho 35, 61, 72
  progressive 73
  serial 72
descending reflectivity core (DRC) 186
deviant motion 37
diagnosis 6, 12
diagnostic parameter 8
differential phase shift 173
differential reflectivity 172
dipole, thunderstorm

115, 118
discrete initiation
  ahead of MCS 75, 177
dissipating stage 28
downburst 35
  dry 36
downdraft 28, 33
  and MCS 62
Downdraft Convective Availability of Potential Energy (DCAPE) 141
downshear 20
downslope 215
downwind 18
DPE. See dynamic pipe effect
DRC. See descending reflectivity core (DRC)
dry adiabat. See adiabat: dry
dry growth 102
dryline 217, 222
dual doppler radar 175
dual-polarization radar 169
dual scan 165
dust devil 91
dynamical models 7
dynamic pipe effect 85

E

effective bulk shear. See shear: effective bulk
effective Bunkers method 150
effective inflow layer. See parcel: effective layer
effective internal dynamics (ID) method 150
efficiency
  of precipitation 52, 164
EF scale. See Enhanced Fujita scale
EHI. See Energy-Helicity Index (EHI)
electrostatic charge 113
electrostatic effect 116
elevated convection. See convection: elevated
Elevated Mixed Layer (EML) 138, 217
emagram 127
EML. See Elevated Mixed Layer (EML)
Energy-Helicity Index (EHI) 246
Energy Shear Index 107
enhanced cumulus 198
Enhanced Fujita scale 93
Enhanced Steering Potential (ESP) 248
enhanced-V 200
entrainment 29, 137
equilibrium level 29
equivalent potential temperature 155
equivalent potential vorticity (EPV) 156
ESP. See Enhanced Steering Potential (ESP)
evaporation 27
evapotranspiration 209

F

fair weather field 114
Fawbush & Miller method 105
FFD. See forward flank downdraft
field mill 120
fine lines 187
Finley, John 3
flanking line 30
flooding 51
flowchart 9
forecast funnel 10
forecasting 3, 12
forward flank downdraft 44
Foster & Bates method 106
front 213
Fujita 1981 scale 16
Fujita damage scale 92
Fujita, Theodore 5, 44

G

geopotential height 203
geostationary 191
Geostationary Lightning Mapper (GLM) 120
geostrophic absolute momentum 155
GOES 191
graupel 101, 117, 119
ground-relative wind 144
gustnado 90

H

hail 101
  formation 101
Hail Detection Algorithm (HDA) 181
hand analysis 10
haze 231
heat burst 34
helical striations 32
helicity 20
  Storm-Relative Helicity (SRH) 152
Herlofson, Nicolai 127
high-based convection. See convection: high-based
hodograph 144
  curved 146
hook echo 46, 184
horizontal shear 19
horizontal vorticity 21
HP. See supercell
humidity
  specific 52
Hydrometeor Classification Algorithm (HCA) 174

I

IC. See lightning
ice
  clear 102

rime 102, 117
incus 30
inflow convergence 178
inflow notch 49
infrared 191
ingredient-based forecasting 12
instability
  and shear 153
  baroclinic 127
  gravitational 125
  inertial 125
  potential 126, 135
  symmetric 126
internal dynamics (ID) method 149
inverted-V 36
isallobaric analysis 212
isentropic lift 223

J

jet stream 206
Johns, Robert 72

K

kidney bean 49
Klemp, Joseph 38

L

lapse rate 137
LCL. See Lifted Condensation Level (LCL)
leading edge mesoscale vortex 72
leading inflow jet 67
leading stratiform. See stratiform precipitation area
Level of Free Convection (LFC) 132
LEWP. See line echo wave pattern
LFC. See Level of Free Convection (LFC)

LI. See Lifted Index (LI)
Lifted Condensation Level (LCL) 131, 144
Lifted Index (LI) 142, 245
lightning 111
  cloud to air (CA) 112
  cloud to cloud (CC) 112
  cloud to ground (CG) 111
  in cloud (IC) 112
lightning detection 119
line echo wave pattern 179
line-end vortex 71
Llano Estacado 228
longitudinal rolls 196
long wave 204
low-level jet (LLJ) 207
LP. See supercell

M

Magnus theory 38
mammatus 55
MARC. See Mid-Altitude Radial Convergence (MARC)
mature stage 28
MCS Index 248
MCV. See mesoscale convective vortex (MCV)
media 97
Melting Layer Detection Algorithm (MLDA) 174
mesocyclone 32, 178
mesohigh 212
mesolow 212
mesoscale 15, 16
mesoscale convective system (MCS) 59
  and lightning 119
mesoscale convective vortex (MCV) 67
mesoscale meteorology 5
meteorological cancer 12
MFC 210
microburst 35
microscale 15
Mid-Altitude Radial Convergence (MARC) 179
Miller, Robert 4
minisupercell. See supercell
misoscale 16
Mixed Layer (ML) parcel. See parcel: Mixed Layer (ML)
mixing ratio 130
  mean 133
  saturation 130
modification, sounding. See sounding: modification of
moist symmetric instability (MSI) 154
moisture 207
moisture convergence 209
moisture flux convergence 210
momentum 155
Most Unstable (MU) parcel. See parcel: Most Unstable (MU)
mothership 47
motion
  deviant 37
  storm 37
MSI. See moist symmetric instability (MSI)
multicell 42
  cluster 43
  line 43
multicell convective cluster (MCC) 61
multispectral imagery 191

N

negative charge 114
negative screening layer 114
NEXRAD 162
North American monsoon 229
northwesterly flow 227
Nyquist velocity 165

O

objective methods 6
occlusion 49
   downdraft 87
open cell 195
ordinary thunderstorm 42
Orlanski scale 15
outflow 34, 216
outflow boundaries 198
outflow boundary 223
overshooting top 198

P

parallel stratiform. *See* stratiform precipitation area
parcel
   effective layer 134
   Mixed Layer (ML) 132
   Most Unstable (MU) 133
   Surface Based (SB) 132
pedestal cloud 53
perturbations, pressure 40
PFJ 206
planetary scale 16
plow winds 46
polar front jet 206
polarimetric radar 169
polarity 121
polarization diversity radar 169
pooling 210
positive area 137
positive charge 114
potential symmetric instability (PSI) 155
potential temperature 215. *See* temperature: potential
pre-squall low 67
pressure fall wave 212
prognosis 6
propagation 37
   MCS 74
PSI. *See* potential symmetric instability (PSI)
pulse repetition frequency 165
pulse thunderstorm 41

Q

quadrant 206
Quantitative Precipitation Estimate (QPE) 174
quasi-geostrophic disturbance 204
quasi-linear convective system (QLCS) 60

R

radar 161
   storm evolution 175
radiance 192
rain 27
rainfall 51
range folding 164
Raton Ridge convergence vorticity zone 229
rear flank downdraft 44
   warmth 94
rear inflow 66
rear-inflow jet 67, 70
rear inflow notch 179
reflectivity 176, 180
reflectivity gradient 176
refraction 168
Renick and Maxwell method 106
return flow 208
RFD. *See* rear flank downdraft
right-hand rule 22
RIJ. *See* rear-inflow jet
rime ice. *See* ice
RIN. *See* rear inflow notch
RKW theory 68, 69
rocket lightning 112
roll cloud 55

S

satellite imagery 191
saturated equivalent potential temperature 155
scales of motion 14
SCAPE. *See* slantwise convective availability of potential energy (SCAPE)
SCP. *See* Supercell Composite Parameter (SCP)
scud cloud 55
sea-level pressure (SLP) 211
SELS 4
Serranias del Burro 229
severe 41
shear 19
   and hodograph 146
   and instability 153
   bulk 146, 147, 149
   cumulative 146
   depth 152
   effective bulk 148
   horizontal 19
   total 146
   vertical 19
shear depth 152
shear vector 19. *See* vector: shear
shelf cloud 54
SHIP. *See* Significant Hail Parameter (SHIP)
Showalter Stability Index (SI, SSI) 141, 245
Significant Hail Parameter (SHIP) 107
Significant Tornado Parameter (STP) 247
skew T log P diagram 127. *See also* sounding
SKYWARN 96
slantwise convection. *See* moist symmetric instability (MSI)
slantwise convective availability of potential energy (SCAPE) 156
smoke 232
Snellman, Leonard 12
sounding 127

modification of 134
parcel 136
proximity 136
southeast U.S. supercell 49
southwest monsoon 229
specific differential phase 173
spectrum width 162, 182
speed shear 20
splitting 39
spotter 54
squall line 4
SRH. See helicity
SSI. See Showalter Stability Index (SI, SSI)
steering flow 149
STJ 206
storm cyclone 79
storm motion 37, 51, 149
storm-relative wind 153
  anvil 49
storm scale 17
storm top divergence 178, 183
STP. See Significant Tornado Parameter (STP)
stratiform precipitation area 62
  and lightning 119
stratocumulus 29, 194
streamwise vorticity 21, 33
Stüve diagram 127
subjective methods 10
subrefraction 168
subtropical jet 206
supercell 43
  anticyclonic 51
  classic (CL) 43, 46
  high-precipitation (HP) 43, 49
  low-precipitation (LP) 43, 47
  low-topped 50
  mini 50, 148
Supercell Composite Parameter (SCP) 247
supercell echo appendage 184

supercooled water 101
superrefraction 168
Surface Based (SB) parcel. See parcel: Surface Based (SB)
synoptic scale 15, 16

T

TBSS. See three body scatter spike (TBSS)
temperature
  potential 129
tephigram 127
theta 129
theta-e 129
three body scatter spike (TBSS) 180
thunderstorm 27
  radar evolution 175
thunderstorm dipole 115
Thunderstorm Project 27, 59
thunderstorms 3
thunderstorm tripole 116
Tornadic Vortex Signature (TVS) 185
tornado 79
  cyclic 88
  life cycle 80
  line-end vortex 89
  QLCS 88
  spinup 85
tornado cyclone 32, 79
Tornado Detection Algorithm (TDA) 185
tornadogenesis. See tornado
tornado scale 17
total shear. See shear: total
trailing stratiform (TS). See stratiform precipitation area
training 52
transverse rolls 195
triple point 222
tripole 116
tuba 80

TVS. See Tornadic Vortex Signature (TVS)

U

unicell 42
unidirectional 20
updraft 28, 29
upflow 65
upshear 20
upslope 223
upwind 18

V

vector
  shear 147
  wind 147
veering 18
Vertically Integrated Liquid (VIL) 183
vertical shear 19
Vertical Totals (VT) index 137
vertical vorticity 21
VIL density 183
visible 191
volume scan 161
vortex
  tornado. See tornado
vorticity 21
  absolute 21
  baroclinic and MCS 68
  crosswise 150
  horizontal 21
  horizontal and MCS 68
  relative 21
  streamwise 150
  vertical 21
VT. See Vertical Totals (VT) index

W

wake low 67
wall clouds 53
warm front 213, 222
waterspout 90

water vapor 191
waves 204
WBZ. *See* wet-bulb zero
weak echo region (WER) 176, 183
weak symmetric stability (WSS) 157
WER. *See* weak echo region (WER)
wet-bulb zero 104, 106
wet growth 101
Wilhelmson, Robert 38
wind direction 17
wind vector. *See* vector: wind
WSR-88D 162

CPSIA information can be obtained
at www.ICGtesting.com
Printed in the USA
LVOW01s0341051215
465019LV00002B/5/P